1st International Conference on Mechanical Engineering, Automation and Control Systems (MEST 2017)

IOP Conference Series: Materials Science and Engineering Volume 203

Johor Bahru, Malaysia
19 – 20 April 2017

ISBN: 978-1-5108-4181-9
ISSN: 1757-8981

Printed from e-media with permission by:

Curran Associates, Inc.
57 Morehouse Lane
Red Hook, NY 12571

Some format issues inherent in the e-media version may also appear in this print version.

Printed by Curran Associates, Inc. (2017)

For permission requests, please contact the Institute of Physics
at the address below.

Institute of Physics
Dirac House, Temple Back
Bristol BS1 6BE UK

Phone: 44 1 17 929 7481
Fax: 44 1 17 920 0979

techtracking@iop.org

Additional copies of this publication are available from:

Curran Associates, Inc.
57 Morehouse Lane
Red Hook, NY 12571 USA
Phone: 845-758-0400
Fax: 845-758-2633
Email: curran@proceedings.com
Web: www.proceedings.com

TABLE OF CONTENTS

IOP Conf. Series: Materials Science and Engineering **203** (2017) 011001 doi:10.1088/1757-899X/203/1/011001

MEST 2017 Chairman's Preface

Bismillahi Rahmanir Rahim – In the name of Allah the Most Gracious and the Most Merciful for His blessings to all of us for having successfully establishing The 1st International Conference in Mechanical Engineering, Science and Technology (MEST 2017).

The 1st MEST2017 was successfully organized by Tensei Innovative Solutions and Metro Polytechnic (Politeknik Metro) Johor Bahru in conjunctions of bridging the connections between the Malaysian's small and medium enterprises (SME) and the technological and research activities within the higher institutions. Undeniably the SME's in Malaysia are in dire need to collaborate their business activities with the higher institution researches in upbringing their entrepreneurship and businesses to higher levels.

A heartily welcome to all our participants and delegates coming from various institutions both locally and internationally. We have participants submitting their papers, coming afar as the Kingdom of Saudi Arabia, Iraq and Iran. And not forgetting from the Malaysians higher institutions as well.

MEST2017 would like to express our highest gratitude to our honorable invited keynote speakers; Assoc. Professor Dr. Abdullah Saleh Alsuwaiyan from the Kingdom of Saudi Arabia, Assoc. Professor Dr. Waluyo Adi Siswanto from JogJakarta, Indonesia and Dr. Abdul Latif Mohd Tobi from UTHM for finding the time and sharing their wonderful researches and experiences with us.

We at Tensei Innovative Solutions whom we considered experts in creative problem solving in customized machineries and consultancies are very grateful for the supports from the participants and industrialists for their contributions to creatively and diversify the knowledge bridging activities. We will strive to keep on publishing peer reviewed papers with Scopus indexed with one purposed only, which is to benefit the graduate students and researchers to keep on producing quality academic papers at an affordable and reasonable expenses and we hope to continue doing so in the future. We deeply hoped that all the postgraduate students could utilize this platform to keep on growing in terms of networking, presentation skills and while sharing their research activities and foreseeing their opportunities to share their research findings and bringing the knowledge gained at their research institutions towards the industries especially the small and young entrepreneurs internationally and locally.

And we sincerely hope to see more and more researchers and students in sharing the knowledge and expanding their networking beyond the wall of higher institutions realistically towards the industries in our next conferences.

Sincerely

Dr. Mohd Sallehuddin Bin Yusof

Chairman MEST2017

IOP Conf. Series: Materials Science and Engineering **203** (2017) 011002 doi:10.1088/1757-899X/203/1/011002

LIST OF COMMITTEE AND EDITORIAL: MEST2017

1. Conference Chair MEST2017

Dr. Mohd Sallehuddin Bin Yusof

Tensei Innovative Solution
No. 51, Blok C, UTHM Kampus Bandar,
Parit Raja, 86400 Batu Pahat,
Johor, MALAYSIA
Email: mdsalleh@mest-ic.org

Faculty of Mechanical and Manufacturing Engineering,
Universiti Tun Hussein Onn Malaysia (UTHM),
86400 Parit Raja, Batu Pahat,
Johor, MALAYSIA
Email: mdsalleh@uthm.edu.my

2. Program Chair MEST2017

Assoc Prof Dr Waluyo Adi Siswanto

Faculty of Mechanical and Manufacturing Engineering,
Universiti Tun Hussein Onn Malaysia (UTHM),
86400 Parit Raja, Batu Pahat,
Johor, MALAYSIA
Email: waluyo@uthm.edu.my

3. Publication Chair MEST2017

Dr. Abdul Latif Mohd Tobi

Tensei Innovative Solution
No. 51, Blok C, UTHM Kampus Bandar,
Parit Raja, 86400 Batu Pahat,
Johor, MALAYSIA
Email: abdlatif@mest-ic.org

Faculty of Engineering Technology,
Universiti Tun Hussein Onn Malaysia (UTHM),
86400 Parit Raja, Batu Pahat,
Johor, MALAYSIA

4. Publicity Chair MEST2017

Dr. Mohammad Zulafif bin Rahim

Tensei Innovative Solution
No. 51, Blok C, UTHM Kampus Bandar,
Parit Raja, 86400 Batu Pahat,
Johor, MALAYSIA
Email: zulafif@mest-ic.org

IOP Conf. Series: Materials Science and Engineering **203** (2017) 011002 doi:10.1088/1757-899X/203/1/011002

Faculty of Mechanical and Manufacturing Engineering,
Universiti Tun Hussein Onn Malaysia (UTHM),
86400 Parit Raja, Batu Pahat,
Johor, MALAYSIA
Email: zulafif@uthm.edu.my

IOP Conf. Series: Materials Science and Engineering **203** (2017) 011002 doi:10.1088/1757-899X/203/1/011002

TECHNICAL EXPERTS: MEST2017

1. Dr. Abdul Latif Mohd Tobi
2. Dr. Affero Ismail
3. Dr. Ahmad Salahuddin Mohd Harituddin
4. Mr. Atheer Hashim
5. Dr. Mohd Yazid Bin Abu
6. Dr. Muhammad Aizzat Bin Zakaria
7. Dr. Hanida Abdul Aziz
8. Dr. Hasyamuddin Bin Othman
9. Mr. Helmy Bin Mustafa El Bakr
10. Dr. Husnul Azan Tajarudin
11. Dr. Mohamad Aizat Abas
12. Mr. Mohamad Ridzuan Bin Mohamad Kamal
13. Dr. Mohammad Zulafif Bin Rahim
14. Dr. Mohd Aizudin Bin Abd Aziz
15. Dr. Mohd Erfy Ismail
16. Dr. Mohd Fairus Mohd Yasin
17. Mr. Mohd Nadzeri Omar
18. Dr. Mohd Sallehuddin Bin Yusof
19. Dr. Muhammad Fairuz Remeli
20. Dr. Nor Dalila Nor Affandi
21. Ir. Dr. Sarjito
22. Dr. Sivakumar A/L Dhar Malingam
23. Dr. Supriyono
24. Assoc. Prof. Dr. Waluyo Adi Siswanto
25. Assoc. Prof. Dr. Wan Fauziah Wan Yusoff
26. Dr. Zainon Binti Mat Sharif

IOP Conf. Series: Materials Science and Engineering **203** (2017) 011003 doi:10.1088/1757-899X/203/1/011003

Peer review statement

All papers published in this volume of *IOP Conference Series: Materials Science and Engineering* have been peer reviewed through processes administered by the proceedings Editors. Reviews were conducted by expert referees to the professional and scientific standards expected of a proceedings journal published by IOP Publishing.

IOP Conf. Series: Materials Science and Engineering **203** (2017) 012001 doi:10.1088/1757-899X/203/1/012001

Designation of Soap Molder Machine and Procedure for Transparent Soap

Zainon Binti Mat Sharif[1,*], Norhasnina Binti Mohd Taib[1], Mohd Sallehuddin Bin Yusof[2], Mohammad Zulafif Bin Rahim[2], Abdul Latif Bin Mohd Tobi[3], Mohd Syafiq Bin Othman[4]

[1]Graduate Business School, College of Graduate Studies, Universiti Tenaga Nasional, Putrajaya Campus, Jalan IKRAM-UNITEN, Kajang, Selangor, Malaysia.

[2]Faculty of Mechanical and Manufacturing Engineering, University Tun Hussein Onn Malaysia (UTHM) P.O. Box 101, Parit Raja, Batu Pahat 86400, Johor, Malaysia.

[3]Faculty of Engineering Technology, University Tun Hussein Onn Malaysia (UTHM) P.O. Box 101, Parit Raja, Batu Pahat 86400, Johor, Malaysia.

[4]Faculty Hotel and Tourism, Universiti Teknologi Mara, Bertam Campus, 13200 Pulau Pinang, Penang, Malaysia.

*zainon@uniten.edu.my

Abstract. Transparent soap is actually the combination of actual soap and solvent. The solvent is added into the soap solution to produce the transparent characteristic. The problem from the previous production is that tiny air bubbles were observed inside the soap resulted in less attractive appearance. Current method of producing the soap bar had taken more than 8 hours and having difficulties to take out the soap bar from the plastic mold with low production rate. It is expected that the air bubble problem can be solved using this new soup molder machine. The soap production rate is believed to increase with the invention of soap molder machine. By reducing the production time from 8 hours to 2 hours, it improve production rate significantly.

Keywords: Soap; air bubble; soap molder machine; transparent soap

1. Introduction

Soap is defined as the alkali salt of fatty acids and was discovered accidentally thousands years ago. According to Roman legend, the soap was first discovered near Mount Sapo, a location near Rome where animal sacrifice was conducted [1]. Initially, the chemical used in a soap making process are animal fat, mixed with the wood ashes as a source of alkali, and rainwater which resulted in a strange yellow mixture. Roman housewives found out that their cloth become cleaner and brighter when using the yellow mixture compared to ordinary water.

Soap is produced by the saponification process, by combining triglycerides (fats and oils) or fatty acids with alkali. According to Palmaist et al [2], saponification reaction is considered as an auto-catalytic since the soap produced is capable of dispersing neutral fatty acid into a colloidal suspension and also capable of dissolving lye. The selection of alkali material depends on the final product solubility and hardness. Two common alkali materials used are sodium and potassium where sodium is used to produce a hard soap while the potassium is used in liquid soap due to its greater solubility. The characteristics of triglycerides and fatty acids also play an important role not only in the features and performance of the final product, but also in its production cost. Commercial triglyceride used in soap production consists of plant source (palm kernel, oils of coconut, olive, etc) and of animal origin (fats of tallow, lard, etc) [1].

Transparent soap is actually the combination of actual soap and solvent. The solvent is added into the soap solution to produce the transparent characteristics. Transparent soap is created by the semi boiled process where the complete mixing and the hardening period only take a few hours. Initially the soap is made using a hot process method where lye (sodium hydroxide), water and oils are mixed together. However, the combination of fatty acids and alkali will produce a soap that is opaque as the fatty acid crystal do not allow light to pass them as a result of light reflection by the crystal. In the production of transparent soap, a solvent consists of one or more substances which act as retarders will be added to reduce the formation of crystals [3]. Mabrouk [4] recorded a few substances that can be used to reduce the formation of crystal in the transparent soap making process namely castor oil, sugar, glycerine and ethanol. Richardson [5] enlarge the list to include cane sugar or sorbitol, glycerol, ethyl-methyl alcohols, and alkali-metal salts of rosin. The solvent was added to dissolve the soap until the crystal formed are so small that the light can freely pass the soap, which renders the soap to be transparent. The solvent used may vary depends on the available material and price.

Several steps are needed to produce the transparent soap where the processes can be quite tedious. However, there have been companies that produce and sell transparent-soap-base commercially. The base has the ability to be fabricated, cooled, melted and re-cooled again and make it convenient to the soap makers. Even though the market-readied transparent soap base has high transparency, yet handling the transparent base quite difficult in the final stage. It also yields tiny air bubbles inside the soap and make it less attractive as can be seen in Figure 1. The problem had caused huge losses in the sale since buyers are not interested and dissatisfied with the product appearance. Another problem is the low production rate since the production was done using a lab scale PVC pipe that only allows few soap bars are produced per cycle. Apart from that, the current method of producing the soap bar had taken more than 8 hours and having difficulties to take out the soap bar from the plastic mold. This paper attempt to discuss the proposed procedure to overcome the air bubble problem and also to design specific equipment that allows higher rate of soap production rate in shorter time. The authors had developed a relatively simple and faster transparent soap production method as discussed below.

Figure 1: Tiny air bubble observed inside the soap

2. Experimental

2.1. Preparation of transparent toilet soap base

Transparent soap base materials use are lye solution, blended oil, solvent and sugar solution. First, the blended oil is heated using a slow cooker until the temperature reaches between 57°C - 62°C. Then, sodium hydroxide is mixed with water inside another container to create a lye solution. For safety purpose, always add the sodium hydroxide gradually into the water. Then, the lye solution is added into the container when the temperature of lye solution almost same with the blended oil. The solution is stirred gently until a thin emulsion was formed. The mixed solution is allowed to cook for almost 2 hours for saponification process. After that, solvent and sugar solution are added and allow to cook for almost 30 minutes or until all the solution completely dissolved. The durations required depend on the amount of the base that wanted to produce. The soap base is then poured into a pipe mold (450 millimetre of length with 40 millimetre in diameter) and cool for 24 hours in room temperature (as shown in Figure 2). The pipe is slanted to about 30° as to provide smooth liquid flows (Figure 3).

2.2 Procedure

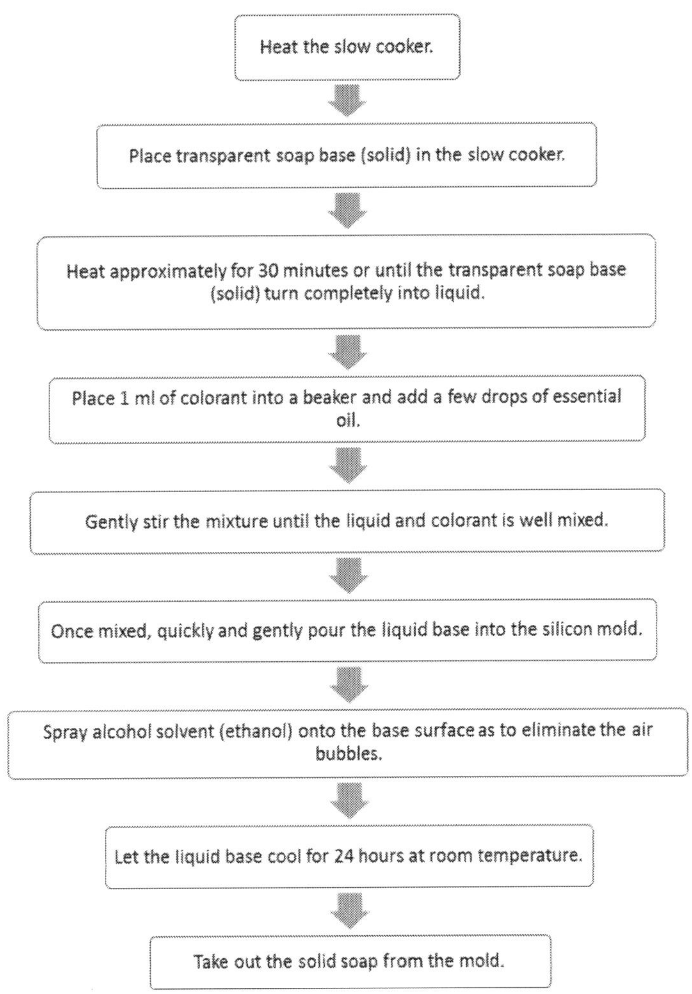

Figure 2: Flowchart

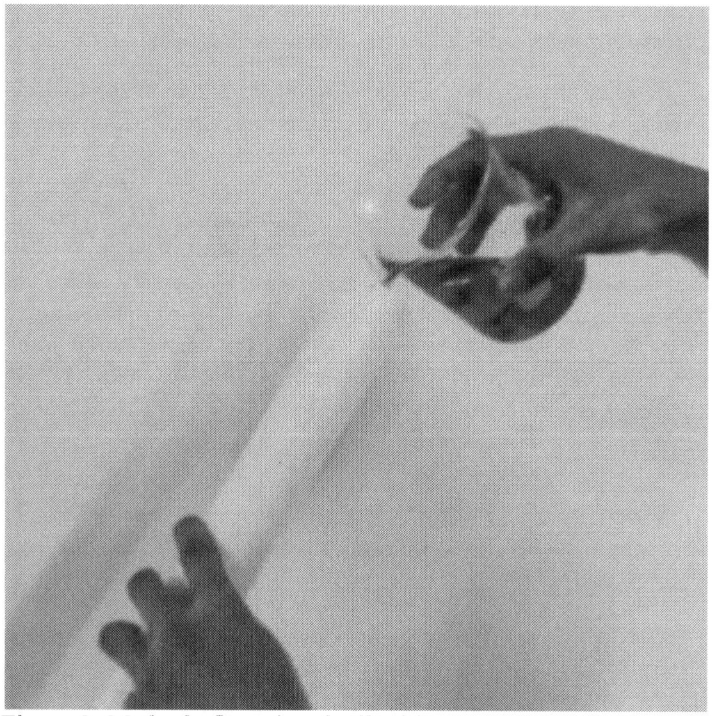

Figure 3: Method of pouring the liquid soap into the plastic mold

Once the soap bars cooled, it was wrapped to prevent moisture exposure and stored for 2 weeks. This is to stabilize the chemical reaction inside the soap. Once it is completed, the transparent soap base can be used for re-melt and pouring process and proceed with the addition of colour and fragrance for the final touch. At this stage, the occurrence of tiny air bubbles is unlikely to happen. If it happened, it is still not affecting the product, since the soap base is not the final product.

2.3 Preparation of transparent soap using soap molder machine

This soap molder machine was designed to increase the production rate and to meet the demand. The machine was designed based on the earlier lab scale method and procedures. The plastic mold is replaced with the aluminium mold with the same height and diameter. The machine consists of two parts where the upper part is the mold and lower part is a mechanical system that able to release the soap from the mold by pushing mechanism. The process involved are discussed below. First the transparent soap base is heated in a slow cooker until temperature reaches 57°C - 62°C. Next the colorant and essential oil is added to the soap base gradually. The amount of colorant and essential oil depends on the market demand and also cost of operation (as shown in figure 4, figure 5 and figure 6).

IOP Conf. Series: Materials Science and Engineering **203** (2017) 012001 doi:10.1088/1757-899X/203/1/012001

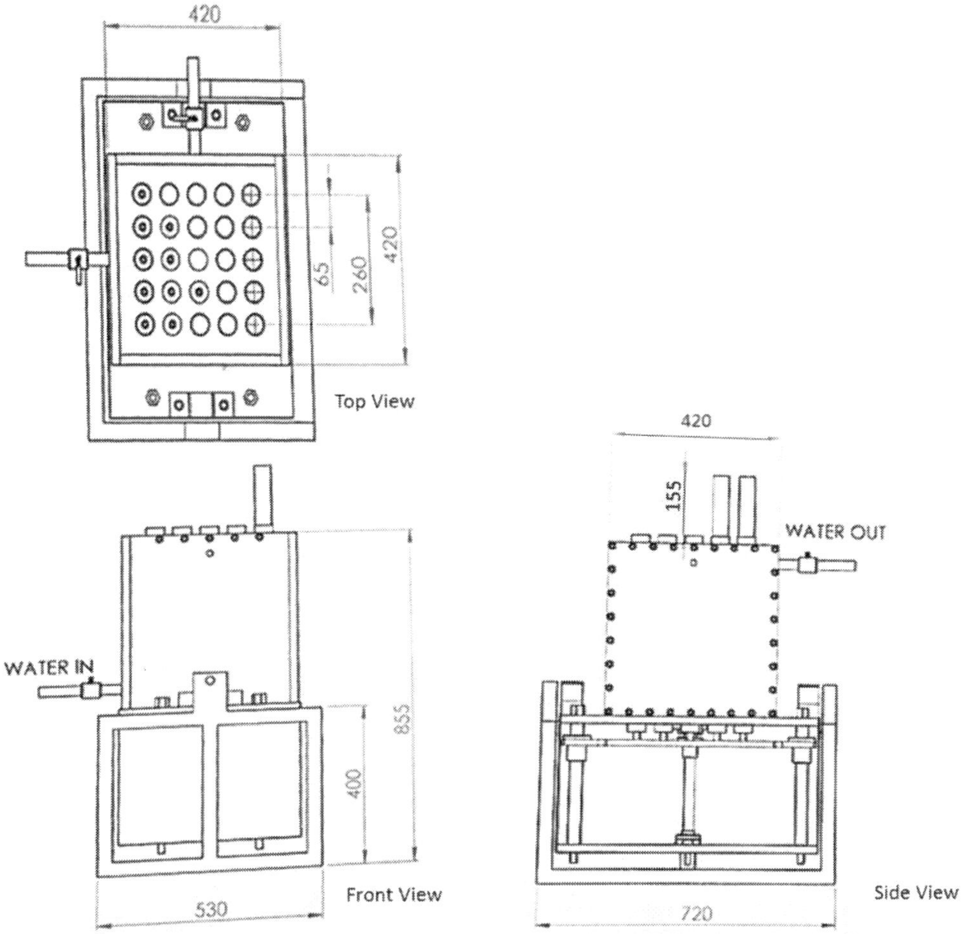

Figure 4: Layout diagram of the soap molder machine

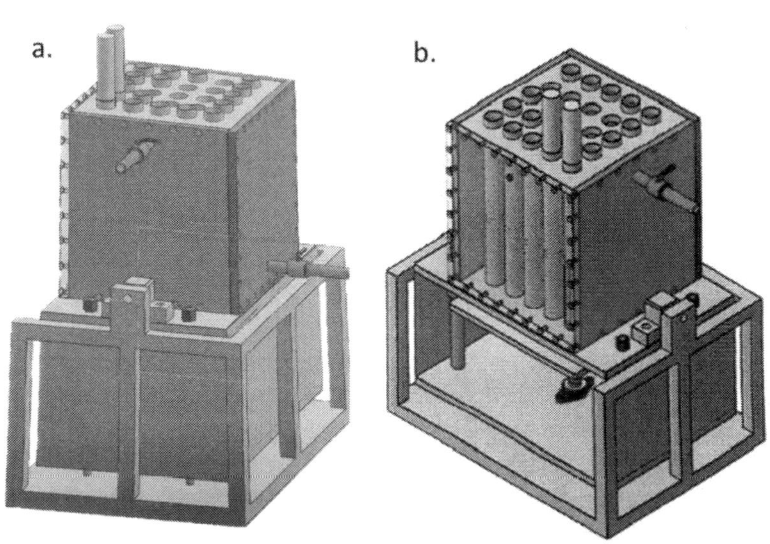

Figure 5: Diagram of the soap molder machine; a. Isometric View b. Sectioning View

IOP Conf. Series: Materials Science and Engineering **203** (2017) 012001 doi:10.1088/1757-899X/203/1/012001

Once all the substances are well mixed, the soap is poured into the mold. This molder machine has 25 aluminium molds (5×5) as depicted in Figure 3. During the pouring stage, the mold part can be adjusted to tilt up to 30° to have the same pouring mechanism as Figure 2. The soap mixture is allowed to cool for about 1 hour at room temperature.

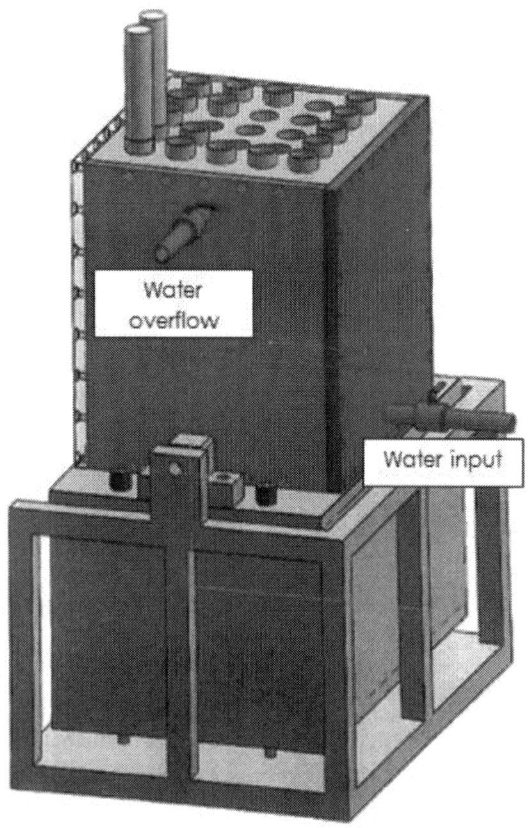

Figure 6: Location of water input and outflow

The cooling process involved passing the cool water into the mold part through the pipe located at the side of the upper part as shown in Figure 5. The temperature of the water ranges between 10°C to 16°C in order to increase the cooling rate of the soap bars. The water will be left for about 1 hour before the soap can be removed from the molder machine.

3. Experimental Result

Observation on the soap produced clearly show no presence of air bubble. The laboratory experience indicate that it is important to ensure the temperature of the transparent soap base is heated in the suggested range before it is poured into the mold. The arrangement of the aluminium mold is critical in order to promote good heat transfer in each mold. It also allows faster cooling period for the soap. This in turn provide enough time for the air bubble produced during the pouring stage to float and moves up to the surface. The slanted pouring method also helps in reducing the liquid turbulence in order to minimize the air bubble formation. This soap production rate is believed to increase with the invention of this soap molder machine. By increasing the numbers of transparent soap molder machine in operation, it will produce a larger number of soap bars. While the reduction in production time from 8 hours to only 2 hours will absolutely have the positive impact on the production rate.

4. Conclusion

The production of high quality transparent soap can be achieved via controlling the temperature and soap base mixture. A new transparent soap molder machine was design to produce transparent soap. Experimental result shows that the soap produced using this machine is transparent and solved the air bubble problem. Furthermore, this machine allows the soap production rate to reduce from about 8 hours to 2 hours.

Acknowledgement

The authors would like to thank University Tenaga Nasional (UNITEN) for giving the permission to publish this paper.

References

[1] Friedman, M., Wolf, R. (1996). Chemistry of soaps and detergents: various types of commercial products and their ingredients. *Clinics in Dermatology,* 14:7-13, SSDI 0738-081X(95)00102-6
[2] Palmqvist, Fredrik, T, E., Sullivant, Frank, E.. (1959). A new approach to continuous soap making-constant composition control. *J. Am. Chem. Soc.* Vol 36, 173-178
[3] Saxton, K., Crosby, B., Dunn, K.. (2013). Formulation of transparent melt and pour soaps without petroleum derivatives. Journal of Sciences. Vol. II
[4] Mabrouk, Suzanne, T,.(2005) Making usable, quality opaque or transparent soap. J Chem Education, Vol 82, No.10, 1534-1537
[5] Richardson, W, D., (1908). Transparent soap – A super cooled solution. 414-420

IOP Conf. Series: Materials Science and Engineering **203** (2017) 012002 doi:10.1088/1757-899X/203/1/012002

A Brief Research Review for Improvement Methods the Wettability between Ceramic Reinforcement Particulate and Aluminium Matrix Composites

Alaa Mohammed Razzaq[1, 4], Dayang Laila Abang Abdul Majid[1], M.R. Ishak[1], Uday M. B[2,3]

[1]Department of Aerospace Engineering, Faculty of Engineering, University Putra Malaysia, 43400 UPM Serdang-Selangor Darul Ehsan-Malaysia. [2]Faculty of Mechanical Engineering, University technology Malaysia, Skudai, 81310 Johor Bahru, Malaysia. [3]UTM Centre for Low Carbon Transport in Cooperation with Imperial College London, Institute for Vehicle Systems and Engineering, University Technology Malaysia, Skudai, 81310 Johor, Malaysia. [4]Oil Production Distribution Company, Ministry Of Oil, Iraqi.

Alaa Mohammed Razzaq; alaaupm@gmail.com, Dayang Laila Abang Abdul Majid; dlaila@upm.edu.my , M.R. Ishak; mohdridzwan@upm.edu.my , Uday M. B; uday@utm.my

Abstract. The development of new methods for addition fine ceramic powders to Al aluminium alloy melts, which would lead to more uniform distribution and effective incorporation of the reinforcement particles into the aluminium matrix alloy. Recently the materials engineering research has moved to composite materials from monolithic, adapting to the global need for lightweight, low cost, quality, and high performance advanced materials. Among the different methods, stir casting is one of the simplest ways of making aluminium matrix composites. However, it suffers from poor distribution and combination of the reinforcement ceramic particles in the metal matrix. These problems become significantly effect to reduce reinforcement size, more agglomeration and tendency with less wettability for the ceramic particles in the melt process. Many researchers have carried out different studies on the wettability between the metal matrix and dispersion phase, which includes added wettability agents, fluxes, preheating the reinforcement particles, coating the reinforcement particles, and use composting techniques. The enhancement of wettability of ceramic particles by the molten matrix alloy and the reinforcement particles distribution improvement in the solidified matrix is the main objective for many studies that will be discussed in this paper.

1. Introduction

Stir casting is the cheapest and simplest technique among different the liquid state production routes. The non-uniform distribution of the particulate is the main problem associated with this process due to gravity regulated segregation and poor wettability[1]. Achieving a uniform dispersion of reinforcement within the matrix is of utmost concern in the stir casting process, as this influences the quality and properties of the developed composites. Thus, the wettability affects the interfacial strength at the matrix-reinforcement joint and is also a key factor in evaluating properties of the metallic matrix composites MMCs [2]. Fig. 1 demonstrates stir casting method.

The wettability can be explained as the spreading capacity of a liquid on a solid surface, and also determines the degree of contact between the liquid and the solid. The buoyant migration can be caused by gas layers at the particles surfaces. Mechanical stirring in a semi solid state is preferred over the completely liquid state so as to promote the breakdown of the gas layer and enhance surface tension reduction. Optimal wettability can be achieved by decreasing the liquid-solid interfacial energy at the reinforcement matrix interface, decreasing the surface tension of the liquid matrix alloy, and increasing the surface energies of the solids.

The adhesion between the matrix and the reinforcement phase results in the interfacial bonding. For adhesion to take place in the course of the composites manufacture, the matrix and the reinforcement must be in a close connection with each other. During the developmental stage of composite, the matrix is often in a state wherein it has a high tendency of flowing towards the reinforcement and this behaviour is similar to that of a liquid flow. A key concept in this contact is wettability. Bonding occurs when an intimate contact is established between the matrix and the reinforcement which is due to adequate wetting of the matrix on the reinforcement particles.

Many methods have been proposed to overcome the problem of the poor wettability. Some of these techniques are complex and expensive, and other simple and cheap methods still to be found. In this review paper, the methods of the wettability improvement will be discussed.

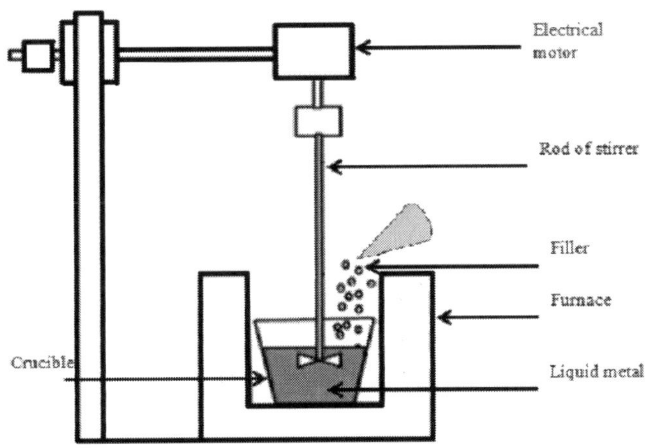

Fig.1. Stir casting method

2. Improvement methods of the wettability

The wettability of a liquid phase on solid substrate can be evaluated using the contact angle measurement of the liquid-solid joint. Meanwhile, the extent of wetting of the interface can be expressed with a contact angle range from 0-180 degree. The contact angles magnitude (θ) describes the wettability in Eq.1[3], as shown in Fig. 2, whereby θ=0 is perfect wettability, $0 < \theta < 180$ is partial wetting, and $\theta = 180$ isn't wetting.

$$\gamma_{gs} = \gamma_{sl} + \gamma_{lg}\cos\theta \quad \text{(Eq.1)}$$

Where $\gamma_{gs}, \gamma_{sl}, and\ \gamma_{lg}$ represent interfacial energies between solid, liquid, and gas phase. In terms of wetting energetic, the adhesion work W_{ad} is the energy required to separate a unit area (J/m^2) of liquid solid interface according to Eq. 2[3].

$$W_{ad} = \gamma_{lg}(1 + \cos\theta) \quad \text{(Eq.2)}$$

The most of engineering methods are used to increase the wettability can be classified into: preheating the reinforcement particles, adding wettability agents, fluxes addition, coating the reinforcement particles, and use compocasting which will be explained in the sections below.

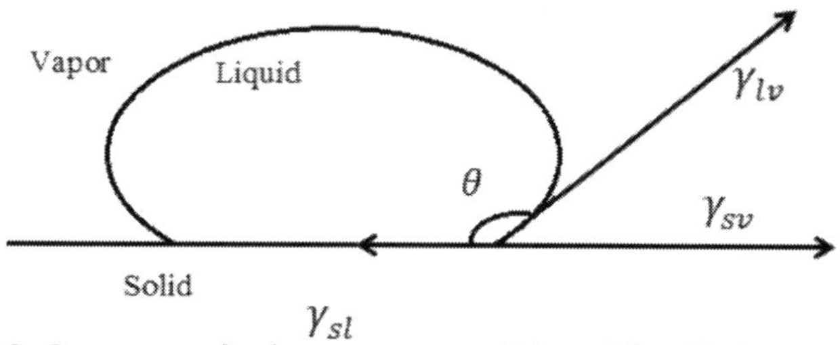

Fig. 2. Contact angles between gas, solid, and liquid phases[4].

2.1 Preheating the reinforcement particles

Essentially, the reinforcements are preheated for two reasons firstly to remove moisture and secondly to make compatible with molten Al. In order to enhance wettability with the liquid alloy or metal matrix material, it is essential to preheat the reinforcement phase.[5]. Heat treatment or calcination of ceramic particles before dispersion into the melt is significant process to get rid of unburnt carbon content, organic materials and release adsorbed gases that will be clear when the powder colour change. Miyajima and Iwai [6], have noticed the preheating of reinforcement particles which used as filler to fabricate composite that improve interfacial strength and the reinforcing particles dispersion. Shirvanimoghaddam et al 2016 studied the influence of processing temperature on the mechanical properties of Al composites which reinforced with micron sized B_4C, $ZrSiO_4$, and TiB_2 by using stir casting. They observed significant improvement 125% increase in the hardness and 52% in tensile strength[7].

2.2 Wettability Agents Addition

Introduction of reactive elements such as Li, Mg, Ca, Ti, Zr and P promotes the wettability of metal-ceramic systems by inducing a chemical reaction at the interface. In addition, reactive elements additions decrease the surface tension of the molten Al and the solid-liquid interfacial energy of the melt [8, 9]. The increase of wettability can be obtained in certain ceramic particles doped with elements having high affinity for oxygen. Sarkar et al [1] reported on the wetting behaviour improvement of two-phase materials using surface properties modification such as the surface tension through the introduction of magnesium to the Al melt. Furthermore, they observed that the ceramic must be subjected to a sieving process so as to prevent having clustered particles, and thereafter, the sieved particles must be pickled and preheated at a certain temperature in order to eliminate moisture and other surface contaminants. The researchers found enhancement of the mechanical properties and wear resistance in the metal matrix compos.

2.3 Fluxes Addition

Potassium fluotitanate K_2TiF_6 flux is often utilized for the purpose of improving the wetting process between Al and Boron carbide B_4C as well as to promote the doping of molten Al with the B_4C particles[10]. They further reported that the development of the Al–B_4C composites having a uniform microstructure is feasible using K_2TiF_6 flux by casting method. Dhinakaran and Moorthy showed that the enhancement of the wettability of B_4C particles in the matrix by using K_2TiF_6 flux into the molten metal has been significant. Reaction of the flux with melted interface surface of B_4C particle resulted in the formation of Ti compounds around the B_4C particle surface region. An exothermic reaction is triggered, followed by heat evolution around the interface of the B_4C particle melt [11]. The improvement at the interface was observed by form a reaction layer containing TiC and TiB_2 at the interface which result in wettability increase and interface bonding[12].Kennedy, AR developed LM25 Al composite by reinforcing B_4C particles of average particle size 25 μm and volume fraction of 5 and 10%. The composites were produced by die casting route, to improve the wettability $K2TiF_6$ flux were added to the melt. The results showed that the improvement of mechanical properties such as the tensile and yield stress. This attribute to a strong interfacial bond between the matrix and reinforcement owing to the evolution of Titanium (Ti) layer around the particles due to the addition of flux[13].

2.4 Coating the Reinforcement Particles

Metallic coated ceramic reinforcements such as graphite, titanium dioxide TiO_2, Al oxide Al_2O_3, Silicon carbide SiC exhibit excellent wetting behaviour and enhanced adhesion when related with their uncoated counterpart. Subjecting ceramic reinforcements to coating is highly essential as it guarantees a limited chemical interaction between the reinforcement and matrix as well as improves the interfacial strength integrity. Thus, paving way for an overall mechanical properties upgrade. [14, 15]. Nickel phosphide Ni–P coated silicon nitride particles have been successfully dispersed in Al6061 alloy. Silicon nitride particle was Ni–P coated Prior its addition to a molten metal and this resulted in having an interface clear of impurities from the reaction products and also an excellent bond was formed between matrix alloy and reinforcement[16].

2.5 Compocasting Technique

Due to the high cost of fabrication involved in the utilization of techniques such as stir casting, it becomes imperative to present, an economical technique that can improve the wettability. The modified approach to the stir casting technique is termed compocasting or slurry casting which includes the casting temperature reduction and the addition of ceramic particles when the Al is in a semi-solid state [17]. Several investigators have reported on the wettability enhancement and uniform distribution of ceramic particles in the developed AMCs using compocasting technique as compared to the stir casting technique [18-20]. Elsewhere [21], results showed that the wettability behaviour and distribution of SiC particles are further enhanced by casting in semisolid state (compocasting) rather than in fully liquid state (stir casting). It also increases the hardness and the impact energy of the composites and decreases their porosity.

It was observed ceramic to be floating when add to molten Al due to high surface tension that leads to poor wettability, the buoyant migration is caused by the gas layers that surround the practical's surfaces. It is possible to break the gas layers by using mechanical stirring in semi sold state thereby reducing surface tension and improve the wettability [22]. Selvam et al. 2013, concluded the incorporation of ceramic particles into semi solid Al alloy improved the wettability[23].

3. Conclusions

Previous research studies have confirmed the strengthening of Al alloys with a dispersion of particulates, thereby strongly enhancing their tribological, mechanical, and structural applications. With a view to achieving a metal matrix composite with optimum properties, several factors must be considered, which includes having a uniformly distributed reinforcement material in the molten matrix as well as an improved wettability or bonding between the matrix and reinforcement. One of the intrinsic drawbacks that bedevil the stir casting technique is the poor wettability of the liquid Al on the reinforcement particles, thereby leading to particle segregation. Improvements of wettability can achieve by surfaces energies increase, decrease of the surface tension of the molten alloy and decrease of the liquid –solid interfacial energy at the interface of the reinforcement matrix. In order to broadly classify the existing engineering approaches suitable for wettability improvement, this paper suggests two basic methods. The first is by subjecting the reinforcement phase to a surface modification process and the melting treatment technique is the second. Surface modified reinforcements can be achieved through particulate treatment at high temperature in order to evaluate the gas desorption and oxidation processes at the surface as well as to determine the coating compatibility of particles with additives that promote the reaction with the matrix. Melt treatment technique is often performed in order to enhance reaction at the metal-particulate interface. In order to reduce to the barest minimum any form of reinforcement degradation, the wetting reaction must be constrained during the fabrication.

4. Acknowledgements:
The authors would like to thanks Aerospace Department in Faculty of Engineering at University Putra Malaysia (UPM) and Mechanical Engineering Faculty, UTM-Centre for Low Carbon Transport, Institute for Vehicle Systems and engineering at Technology University Malaysia (UTM) for providing the research facilities. This research work has been supported by the UPM Research Grant (9463100) and the UTM Research FRGS Grant (R.J1300000.7824.4F723).

References

1. Sarkar, S., et al., Studies on Aluminum—Fly-Ash Composite Produced by Impeller Mixing. Journal of reinforced plastics and composites, 2010. **29**(1): p. 144-148.
2. Akbari, M.K., H. Baharvandi, and O. Mirzaee, Nano-sized aluminum oxide reinforced commercial casting A356 alloy matrix: Evaluation of hardness, wear resistance and compressive strength focusing on particle distribution in aluminum matrix. Composites Part B: Engineering, 2013. **52**: p. 262-268.
3. Ibrahim, I., F. Mohamed, and E. Lavernia, Particulate reinforced metal matrix composites—a review. Journal of materials science, 1991. **26**(5): p. 1137-1156.
4. Hashim, J., L. Looney, and M. Hashmi, Metal matrix composites: production by the stir casting method. Journal of Materials Processing Technology, 1999. **92**: p. 1-7.
5. Sahin, Y., Preparation and some properties of SiC particle reinforced aluminium alloy composites. Materials & design, 2003. **24**(8): p. 671-679.
6. Miyajima, T. and Y. Iwai, Effects of reinforcements on sliding wear behavior of aluminum matrix composites. Wear, 2003. **255**(1): p. 606-616.
7. Shirvanimoghaddam, K., et al., Effect of B 4 C, TiB 2 and ZrSiO 4 ceramic particles on mechanical properties of aluminium matrix composites: experimental investigation and predictive modelling. Ceramics International, 2016.
8. Hashim, J., L. Looney, and M. Hashmi, The enhancement of wettability of SiC particles in cast aluminium matrix composites. Journal of Materials Processing Technology, 2001. **119**(1): p. 329-335.
9. Arslan, G. and A. Kalemtas, Processing of silicon carbide–boron carbide–aluminium composites. Journal of the European Ceramic Society, 2009. **29**(3): p. 473-480.
10. Kerti, I. and F. Toptan, Microstructural variations in cast B 4 C-reinforced aluminium matrix composites (AMCs). Materials letters, 2008. **62**(8): p. 1215-1218.
11. Dhinakaran, S. and T. Moorthy2b, Fabrication and Characteristic of Boron Carbide Particulate Reinforced Aluminum Metal Matrix Composites.
12. Toptan, F., A. Kilicarslan, and I. Kerti. The effect of Ti addition on the properties of Al-B4C interface: A microstructural study. in Materials Science Forum. 2010. Trans Tech Publ.
13. Kennedy, A., The microstructure and mechanical properties of Al-Si-B 4 C metal matrix composites. Journal of materials science, 2002. **37**(2): p. 317-323.
14. Klauser, R., et al., The interaction of oxygen and hydrogen on a diamond C (111) surface: a synchrotron radiation photoemission, LEED and AES study. Surface science, 1996. **356**(1-3): p. L410-L416.
15. Asthana, R. and P. Rohatgi, On the melt infiltration of plain and nickel-coated reinforcements with aluminium alloys. Journal of materials science letters, 1993. **12**(6): p. 442-445.

16. Ramesh, C., et al., Microstructure and mechanical properties of Ni–P coated Si 3 N 4 reinforced Al6061 composites. Materials Science and Engineering: A, 2009. **502**(1): p. 99-106.
17. Amirkhanlou, S., et al., High-strength and highly-uniform composites produced by compocasting and cold rolling processes. Materials & Design, 2011. **32**(4): p. 2085-2090.
18. Ceschini, L., G. Minak, and A. Morri, Tensile and fatigue properties of the AA6061/20vol% Al 2 O 3p and AA7005/10vol% Al 2 O 3p composites. Composites Science and Technology, 2006. **66**(2): p. 333-342.
19. Sajjadi, S.A., H. Ezatpour, and M.T. Parizi, Comparison of microstructure and mechanical properties of A356 aluminum alloy/Al 2 O 3 composites fabricated by stir and compo-casting processes. Materials & Design, 2012. **34**: p. 106-111.
20. Amirkhanlou, S. and B. Niroumand, Fabrication and characterization of Al356/SiC p semisolid composites by injecting SiC p containing composite powders. Journal of Materials Processing Technology, 2012. **212**(4): p. 841-847.
21. Amirkhanlou, S. and B. Niroumand, Synthesis and characterization of 356-SiCp composites by stir casting and compocasting methods. Transactions of Nonferrous Metals Society of China, 2010. **20**: p. s788-s793.
22. Shanmughasundaram, P., R. Subramanian, and G. Prabhu, Some studies on aluminium–fly ash composites fabricated by two step stir casting method. European journal of scientific research, 2011. **63**(2): p. 204-218.
23. Selvam, J.D.R., D.R. Smart, and I. Dinaharan, Microstructure and some mechanical properties of fly ash particulate reinforced AA6061 aluminum alloy composites prepared by compocasting. Materials & Design, 2013. **49**: p. 28-34.

Synthesis of Hydroxyapatite through Ultrasound and Calcination Techniques

John O. Akindoyo[a,b*], M. D. H. Beg[a], Suriati Ghazali[a], Edward O. Akindoyo[a], Nitthiyah Jeyaratnam[a]

[a]Faculty of Chemical and Natural Resources Engineering, Universiti Malaysia Pahang, 26300 Gambang, Pahang, Malaysia.
[b]Institute of Materials Engineering, Polymer Technology, Mönchebergstr. - 3, 34125 Kassel, University of Kassel, Germany

*Corresponding author: John O. Akindoyo; email: blessedbode@ymail.com

Abstract. There is a growing demand for hydroxyapatite (HA) especially in medical applications, production of HA which is totally green is however a challenge. In this research, HA was produced from biowaste through ultrasound followed by calcination techniques. Pre-treatment of the biowaste was effectively achieved through the help of ultrasound. After calcination at 950°C, the obtained HA was characterized through Thermogravimetric (TGA) analysis, X-ray diffraction analysis (XRD) and Fourier transform infrared spectroscopy (FTIR). Spectrum of the produced HA was compared with standard HA index. The spectrum is in agreement with the standard HA as confirmed through FTIR, XRD and TGA result. Furthermore, morphological study of the HA through Field emission scanning electron microscope (FESEM) shows almost uniform spherical shape for the HA as expected. Based on the results obtained herein, combining ultrasound with calcination can help to produce pure HA with potential medical applications without the use of any organic solvent.

1. Introduction

Biomaterials based on inorganic calcium orthophosphate are currently being investigated due to their broad range potential application in the field of medicine especially as bone replacements materials. Hydroxyapatite (HA), is in the forefront of these investigations [1, 2]. Specifically, HA has been reported to exhibit highly desirable bioactive, nontoxic, noninflammatory, biocompatible, nonimunogenic and properties. These properties make HA an important candidate for good osteointegration [3-5]. There have been several reports on the production of HA either from synthetic or natural sources. It was however reported that most of the synthetic techniques are either time consuming, complicated or biologically unsafe leading to safety issues as well as high cost of the produced HA [6, 7].

To overcome this challenge, HA is being extracted from natural sources with potential benefits such as economic feasibility, possibility for large scale production and environmental friendliness [8]. Extraction of HA from biowastes is often achieved through ordinary calcination. Other methods which may be used are subcritical extraction and alkaline hydrothermal processes [6]. However, most of the reported researches often use chemical solvents for fat removal, defleshing/cleaning of the waste materials before extraction. On the other hand, it has also been reported that unlike natural bones, the synthetic approach often eliminate carbonate groups from the obtained HA which may negatively influence the biological features of the apatite obtained [6].

Although research on HA synthesis from natural sources and biowastes date back to about 4 decades ago [10], as much as we know the potential of ultrasound technique have not been exploited. Interestingly, ultrasound method has been perceived to be suitable for technically feasible, easy and environment benign treatment of biowaste before extracting HA from them. Ultrasound technique is

Published under licence by IOP Publishing Ltd

well established in separation technology but has been recently observed to exhibit certain features necessary for biowaste treatment [11, 12].

Some potential benefits of ultrasound include better purification with very little or no alkali concentrations, property enhancement, faster reaction rate and economic feasibility [12]. Combining ultrasound treatment with calcination process is therefore thought to be a great idea for an organic solvent free extraction of HA from biowastes. The aim of this research was to combine the potential of ultrasound with calcination process to synthesize environment benign HA from biowaste namely cow bone and to investigate the properties of the produced HA.

2. Materials and Methods

2.1. Bone Preparation and production of hydroxyapatite

The femur of an about 3 years old cow was collected from Jaya Gading, Kuantan, Malaysia. After initial drying in air for 48 h, the bones were oven dried for 8 h at 60 °C. After this, an Ultrasonic Bath operating at power of 9 W was used to agitate the bones in order to remove fats, flesh, and other adhering materials from the bone. This was continued for 2 h at 100 °C in a water medium ensuring a ratio of 1:50 (w/v) for bone and water respectively. An aluminium sponge in 40 °C distilled water was used to wipe the ultrasound treated bones and then left to dry normally for 24 h after which they were oven dried at 105 °C for 8 h. Grinding was carried out with the help of a Germany Retsch ultra centrifugal mill (ZM 200) after which the bone powder was sieved to obtain a particle size of about 450 μm. The HA was then obtained through thermal treatment of the bone powder via calcination process. This was carried out at 950 °C in a laboratory furnace for 3 h. The obtained material was stored for further analysis whereas an untreated sample of the bone powder was also stored for comparison. The flow chart of the experimental procedure is shown in Figure 1.

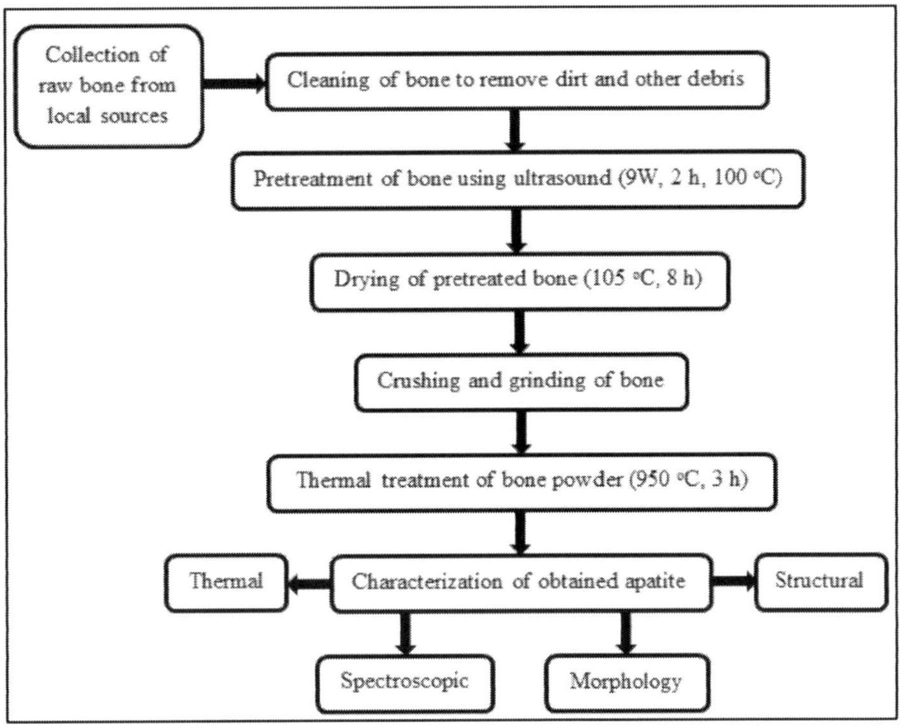

Figure 1. Flow chart of experimental procedure

2.2. Thermogravimetric analysis

Thermogravimetric analysis was conducted on the samples using a TA analyser (TGA Q500 V6.4, Germany). About 5 ± 2 mg samples were placed in a platinum crucible and analysis was carried out under nitrogen atmosphere at a flow rate of 40 mL min^{-1}. This was conducted from room temperature till 1000 °C at 10 °C min^{-1}.

2.3. Fourier Transform Infrared (FTIR) spectroscopy
Functionality of the samples was investigated through FTIR analysis. This was used to study differences in functional groups of the obtained material and the raw bone powder through their spectra analysis. It was also used to identify possible differences between the produced HA and standard HA. The KBr method was used to obtain IR spectra at a wavelength range of 400-4000 cm^{-1}.

2.4. Field emission scanning electron microscopy
Surface morphology of the samples was done through FESEM analysis. FESEM analysis was conducted with a ZEISS, EVO 50 scanning electron microscope. To prevent electrical discharge, the samples were first dried to remove moisture and then sputter coated with platinum prior to FESEM examination.

2.5. X-ray diffraction analysis
X-ray diffraction analysis (XRD) was conducted using an X-ray diffractometer at 1deg min^{-1}over a range of 3-60° at sampling step of 0.02°. For phase identification, spectrum of the obtained material was validated against the XRD spectrum for standard HA (ICCD standard HA, DB card number 01-074-9761).

3. Results and Discussion

3.1. Thermo gravimetric analysis (TGA)
Figure 2 illustrates the TGA spectra for the untreated and treated bone powder. As can be seen from the Figure for the untreated sample, a weight loss occurred from about 50-200 °C which is due to release of adsorbed surface moisture. This stage of weight loss was not significant in the spectrum of the treated sample indicating that the sample is totally dry. Around 250-500 °C the spectrum of the untreated sample showed another stage of weight loss which is attributed to the burning of organic materials present in the raw bone powder. This also was not significant in the spectrum of the treated samples which indicates that organic components have been removed from the treated material. Generally for both samples, a weight loss occurred from 600-800 °C which may be associated with endothermic dissociation of trace CO_3^{2-}.

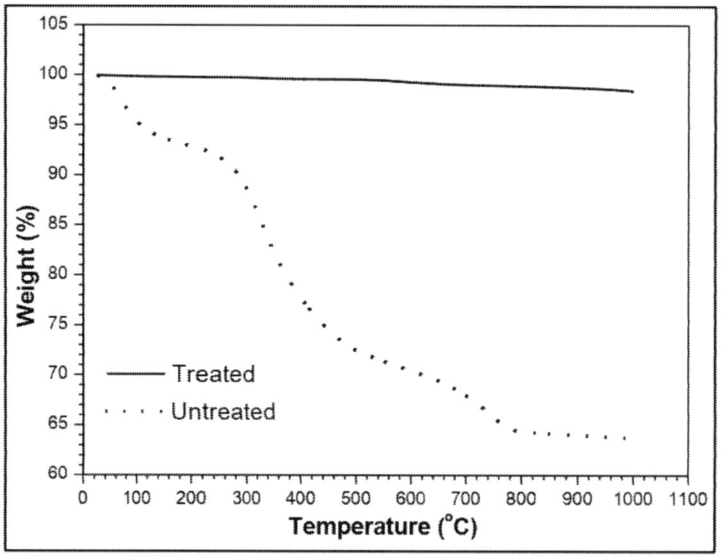

Figure 2. TGA spectra for untreated and ultrasound/calcination treated cow bone powder

It was reported in literature that decomposition of CO_3^{2-} often take place around 400-600 °C in air and 500-890 °C under nitrogen atmosphere [13]. This aligns with the result obtained in this study and it indicates that the synthesized material HA exhibits certain percent of CO_3^{2-} ions. It is worthy of note that on the overall, there was no much pronounced weight loss due to endothermic and exothermic reaction in spectrum of the treated sample. This is an indication of good thermal stability of the obtained material as expected of chemical HA. The overall weight loss for the two samples after 800 °C is 35.5% and 1.0% for the untreated and treated samples respectively. From literature review, it was reported that bone generally consists of about 30-40% organic components and about 60-70% inorganic components [14, 15]. Therefore, the higher amount of weight loss recorded for the untreated sample can be attributed to the presence of larger percent of organic constituents in the untreated bone powder compared with the treated sample.

3.2. Field emission scanning electron microscopy

The FESEM images of the untreated and treated samples are shown in Figure 3. It can be seen from Figure 3 that the produced HA reveals a spherical morphological shape which is nearly uniform and with better dispersion than the untreated sample. HA have been reported to exhibit different shapes with respect to the synthetic method. The spherical shape obtained herein is therefore acceptable [20, 21]. It is interesting to note that the treated sample exhibits reasonable interconnectivity such as those found in natural bone apatites. In addition, it shows great increase in size compared with the untreated sample. This is because HA particles tend to crystallize and grow up at higher temperature as reported in literature [6, 9].

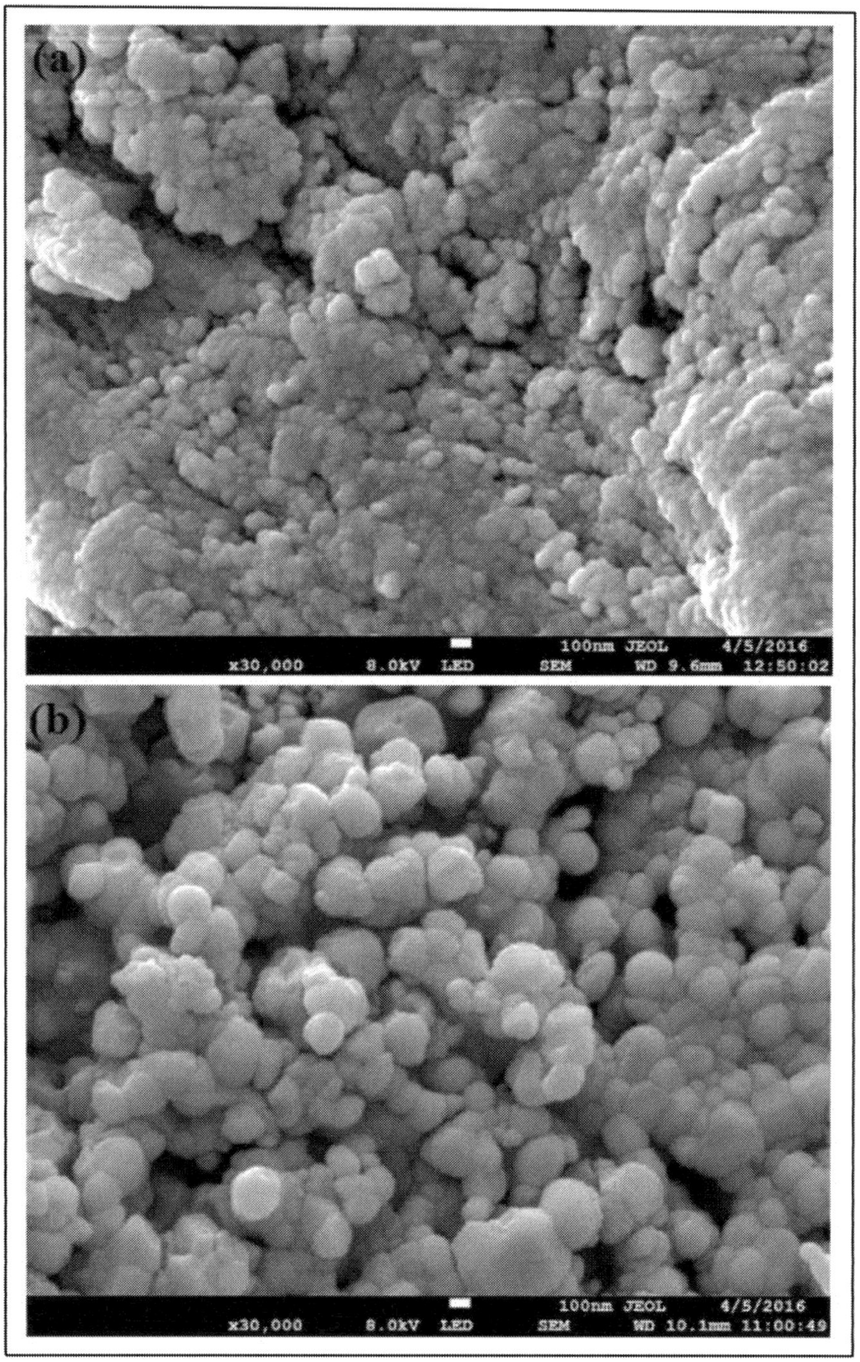

Figure 3. FESEM images for (a) untreated and (b) ultrasound/calcination treated cow bone powder

3.3. Fourier transforms infrared spectroscopy analysis

The FTIR spectra of the untreated and treated samples are illustrated in Figure 4. The band at higher wavelength from 3200-3500 cm^{-1} is an attribute of –OH stretching vibration of bonded hydroxyl groups [11]. This functional group can also be seen at 1637 cm^{-1} in the spectrum of the untreated sample. This indicates the presence of adsorbed surface water on the untreated bone powder [9, 18], as supported by the TGA result of this study. Also, this peak may be attributed to –OH groups from the HA structure [19]. At 2840-2930 cm^{-1} there is a low intense peak which is characteristic of stretching vibrations from C-H bonds [11]. Presence of this band in the spectrum of the untreated sample indicates the presence of organic materials in the untreated bone powder as indicated by the

TGA results. However, this band was not present in the spectrum of the treated sample. This confirms the full removal of organic materials from the bone powder after thermal treatment [19]. The double split peaks which can be seen at 1457 cm^{-1} and 1411 cm^{-1} are characteristics of asymmetric CO_3^{2-} stretching and this shows that CO_3^{2-} was substituted into the lattice of the obtained product. This conforms to the results obtained through TGA and XRD analysis of this study. It is also in agreement with previous research work [18]. The peaks at 1015 cm^{-1} and 960 cm^{-1} are attributed to stretching vibration of phosphate PO_4^{3-} ions whereas the peak around 550-650 cm^{-1} are attributed to phosphate deformational vibrations [7, 9]. The peak at 872 cm^{-1} further indicates the ionic substitution of CO_3^{2-}into the obtained product. This is typical of type –A carbonate substitution, and suggests the presence of surface labile CO_3^{2-} in the material obtained after treatment [9].

Calcium phosphates including HA exhibits a compositional structure which have been reported to be very similar to natural bones [17, 20]. Based on the observations reported herein, it can be concluded that the obtained material is structurally similar to HA and can therefore be exploited for bone replacement purposes.

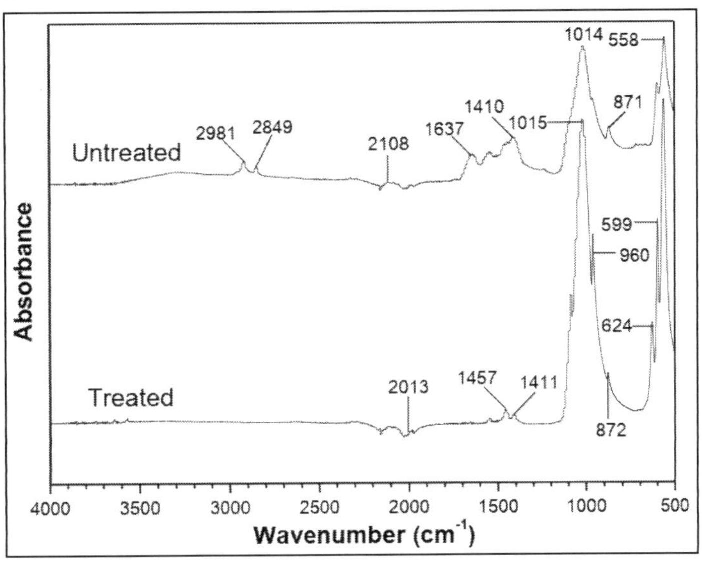

Figure 4. FTIR spectra for untreated and ultrasound/calcination treated cow bone powder

3.4. X-ray diffraction analysis

Figure 5 illustrates the XRD diffractograms for the untreated and treated samples as well as a standard XRD spectrum for HA (ICCD standard HA, DB card number 01-074-9761) in order to validate the synthesized product. As can be seen in Figure 5, XRD pattern of the treated sample is very intense which suggests the crystalline nature of the produced material [16]. The characteristic peaks of standard HA can be seen in the spectrum of the produced material. This indicates that ultrasound treatment followed by calcination was effective to remove collagen and other organic components from the raw bone powder without causing undesired disruption to the molecular skeletal of the obtained HA [21]. Based on the close resemblance of the peaks at 2θ regions 26, 29, 32-34, 40 and 47-54 with standard HA, we can infer that the obtained product is hydroxyapatite (HA) [22].

On the contrary, spectrum of the untreated sample as shown in Figure 5 can be seen to be broader and short of some peaks from the standard HA spectra. This is attributed to the presence of fibrous collagen in the untreated sample which might lead to undesired dispersion of the X-ray radiations [6].

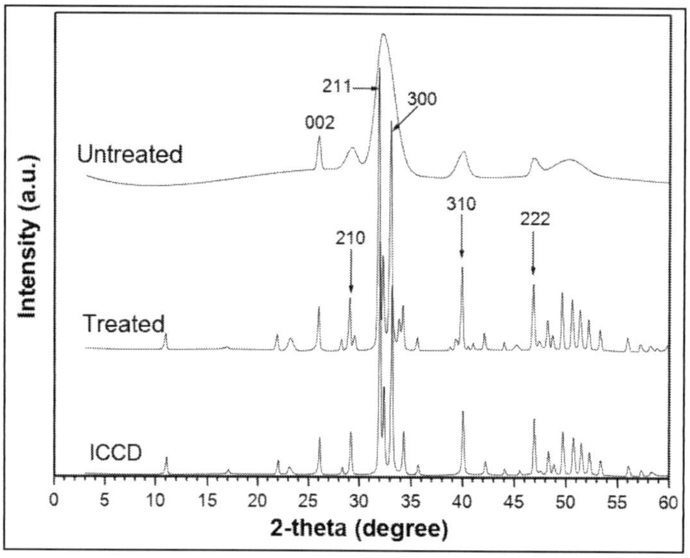

Figure 5. XRD spectra for untreated, ultrasound/calcination treated cow bone powder and standard HA

4. Conclusion

Hydroxyapatite (HA) was extracted from biowaste namely cow bone through a combination of ultrasound treatment in ordinary water medium and calcination process at 950 °C. Treatment with ultrasound was suitable for proper cleaning of the raw cow bone. The HA obtained after calcination at 950 °C for 3 h revealed highly desirable properties in term of morphology, chemical structure and compositional constituents. Results of XRD and FTIR are both consistent and revealed that chemical features of the produced HA conform to the standard HA. Also, there is good agreement in the main phase of the synthesized HA and standard HA as confirmed through all the characterization techniques employed herein. Based on the TGA and FTIR results, it is evident that HA obtained through ultrasound and calcination processes can help to avoid the total elimination of CO_3^{2-} from the HA which will help to enhance the biological properties of the HA. More importantly, the total avoidance of organic solvent through the techniques employed herein can be a way forward on how to reduce the amount of chemicals associated with conventional HA synthesis. Moreover, use of biowaste for HA synthesis can offer both economic and environmental benefits and the HA obtained exhibits desired features for medical application.

Acknowledgement

The authors would like to appreciate Universiti Malaysia Pahang for the financial support through short term grant, RDU1603139

References

[1] Gokcekaya O, Ueda K, Narushima T and Ergun C 2015 Synthesis and characterization of Ag-containing calcium phosphates with various Ca/P ratios *Materials Science and Engineering: C.* **53** 111-119.

[2] He L H, Standard O C, Huang T.T, Latella B A and Swain M V 2008 Mechanical behaviour of porous hydroxyapatite *Acta Biomateriali.* **4** 577-586.

[3] Fu C, Zhang X, Savino K, Gabrys P, Gao P, Chaimayo W and Miller B L 2016 Antimicrobial silver-hydroxyapatite composite coatings through two-stage electrochemical synthesis," *Surface and Coatings Technology.*

[4] Cucuruz A T, Andronescu E, Ficai A, Ilie A and Iordache F 2016 Synthesis and characterization of new composite materials based on poly (methacrylic acid) and hydroxyapatite with applications in dentistry *International journal of pharmaceutics.*

[5] Dorozhkin S V 2015 Calcium orthophosphate deposits: preparation, properties and biomedical

applications *Materials Science and Engineering.* **55** 272-326.
[6] Barakat N A, Khil M S, Omran A, Sheikh F A and Kim H Y 2009 Extraction of pure natural hydroxyapatite from the bovine bones bio waste by three different methods *Journal of materials processing technology.* **209** 3408-3415.
[7] Lu X Y, Fan Y B, Gu D and Cui W 2007 Preparation and characterization of natural hydroxyapatite from animal hard tissues *Key Engineering Materials*. 213-216.
[8] Wu S C, Hsu H C, Hsu S K, Lin F W and Ho W F 2015 Preparation and characterization of porous calcium-phosphate microspheres *Ceramics International.* **41** 7596-7604.
[9] Liu Q, Matinlinna J P, Chen Z, Ning C, Ni G, Pan H and Darvell B W 2015 Effect of thermal treatment on carbonated hydroxyapatite: Morphology, composition, crystal characteristics and solubility *Ceramics International.* **41** 6149-6157.
[10] Champion E 2013 Sintering of calcium phosphate bioceramics *Acta biomaterialia* **9** 5855-5875.
[11] Akindoyo J O, Beg M D H, Ghazali S and Islam M R 2015 Effects of poly (dimethyl siloxane) on the water absorption and natural degradation of poly (lactic acid)/oil-palm empty-fruit-bunch fiber biocomposites *Journal of Applied Polymer Science.* **132**.
[12] Akindoyo J O, Beg M D H, Ghazali S, Islam M R and Mamun A A 2015 Preparation and Characterization of Poly (lactic acid)-Based Composites Reinforced with Poly Dimethyl Siloxane/Ultrasound-Treated Oil Palm Empty Fruit Bunch *Polymer-Plastics Technology and Engineering.* **54** 1321-1333.
[13] Murugan R, Ramakrishna S and Rao K P Nanoporous hydroxy-carbonate apatite scaffold made of natural bone *Materials Letters.* **60** 2844-2847.
[14] Zuber M, Zia F, Zia K M, Tabasum S, Salman M and Sultan N 2015 Collagen based polyurethanes—A review of recent advances and perspective *International Journal of Biological Macromolecules.* **80** 366-374.
[15] Jang D W, Franco R A, Sarkar S K and Lee B T 2014 Fabrication of porous hydroxyapatite scaffolds as artificial bone preform and its biocompatibility evaluation *Asaio Journal.* **60** 216.
[16] Nirmala R, Sheikh F A, Kanjwal M A, Lee J H, Park S J, Navamathavan R and Kim H Y 2011 Synthesis and characterization of bovine femur bone hydroxyapatite containing silver nanoparticles for the biomedical applications *Journal of Nanoparticle Research.* **13** 1917-1927.
[17] Zhou H and Lee J Nanoscale hydroxyapatite particles for bone tissue engineering *Acta Biomaterialia.* **7** 2769-2781.
[18] Liuyun J, Chengdong X, Dongliang C and Lixin J 2012 Effect of n-HA with different surface-modified on the properties of n-HA/PLGA composite *Applied Surface Science.* **259** 72-78.
[19] Haberko K, Bućko M M, Brzezińska-Miecznik J, Haberko M, Mozgawa W, Panz T, Pyda A and Zarebski J 2006 Natural hydroxyapatite—its behaviour during heat treatment *Journal of the European Ceramic Society.* **26** 537-542.
[20] Fielding G and Bose S 2013 SiO 2 and ZnO dopants in three-dimensionally printed tricalcium phosphate bone tissue engineering scaffolds enhance osteogenesis and angiogenesis in vivo *Acta Biomaterialia.* **9** 9137-9148.
[21] Liu F, Wang R, Cheng Y, Jiang X, Zhang Q and Zhu M 2013 Polymer grafted hydroxyapatite whisker as a filler for dental composite resin with enhanced physical and mechanical properties *Materials Science and Engineering: C.* **33** 4994-5000.
[22] Choi H W, Lee H J, Kim K J, Kim H M and Lee S C 2006 Surface modification of hydroxyapatite nanocrystals by grafting polymers containing phosphonic acid groups *Journal of Colloid and Interface Science.* **304** 277-281.

IOP Conf. Series: Materials Science and Engineering **203** (2017) 012004 doi:10.1088/1757-899X/203/1/012004

Utilization of polyethylene terephthalate (PET) in asphalt pavement: A review

A. F. Ahmad[1], A. R. Razali[1], I. S. M. Razelan[2]

[1]Faculty of Mechanical Engineering, Universiti Malaysia Pahang, 26600 Pekan, Pahang, Malaysia
[2]Faculty of Civil Engineering a and Earth Resources, University Malaysia Pahang, Lebuhraya Tun Razak, 26300 Gambang, Kuantan, Pahang

E-mail: amaninafarhana.ahmad@gmail.com

Abstract. The quantity of plastics used throughout the world is increasing every year. Municipal solid wastes (MSW), manufacturing processes and service industries produce a lot of waste plastic materials. The increasing awareness among consumers about the environment has contributed to the concerns over disposal of generated wastes. The growing number of plastic materials every year and limited landfill conditions causes many alternatives exist for the disposal of plastic waste. This paper provides a summary of the study on the utilization of polyethylene terephthalate (PET) in road construction. Data from researcher show that PET can improve some properties of modified asphalt mixture. Having considered the economic and environmental prudent angles, utilization of PET as an additive to asphalt mixture is suitable to be used for road pavement.

1. Introduction

Plastics are regularly utilized substances which play an essential part in practically every part of our lives. The increasing of plastic waste throughout the world need appropriate end-of-life management. Most amount of plastics can be found in containers and packaging (i.e. bottles, cup, etc.) and also can be found in disposable good (e.g. medical device) and durables (e.g. furniture, building materials, tires, etc.) [1]. Compared to other materials, plastics always be selected because of their properties such as easy processing, low density, good chemical resistance, good mechanical properties, good electrical insulating properties, low cost and good thermal properties. There are two main field directions of plastic application for post-production and post-consumer which are used as material recycling of waste polymer and used as alternative fuel in power plants.

In 2002, the world's annual consumption of plastic material was 204 million tons, but it was increased to 300 million tons in 2013[2]. Furthermore, consumption and production of plastics and polymers are expected to be continuously growing in next year. Table 1 show the plastic production by plastic category in the Malaysia [3].

Table 1. Plastic Production by Plastic Category in The Malaysia [3]

Type of plastic	Plastic Production (%)
High Density Polyethylene (HDPE)	24
Low Density Polyethylene (LDPE)	24
Polyethylene Terephthalate (PET, PETE)	13
Polyvinyl Chloride (PVC)	10
Polypropylene (PP)	13
Polystyrene (PS)	10
Polycarbonate (PC)	3
Acrylonitrilebutadienne-styrene (ABS)	4
Polymethyl Methacrylate (PMMA)	2
Others (acrylic, nylon and epoxy resin)	2

In Malaysia, approximately 0.80 kg/capita to 1.9 kg/capita of the municipal solid waste is generated daily, and this is expected to increase every year [4]. Plastic waste is the third largest waste after putrescible waste and paper [3]. Most of the plastic waste is drinking bottles which are made from PET.

There are two main types of plastic which are thermosetting and thermoplastics. Heating can cure thermosetting plastic. It can melt and shape then stay solid after have solidified. During this process, the chemical reaction is irreversible. Another type of plastic is thermoplastic which this cannot undergo chemical charge in composition and it also reversible. Table 2 shows the examples of thermoplastics and thermosetting materials [5]. Therefore, thermoplastic or thermosetting can be used as an additive to the bituminous mixture.

Table 2. Examples Of Thermosetting And Thermoplastics Materials.

Thermoplastics	Thermosetting
Polyethylene Terephthalate (PET)	Bakelite
Polypropylene (PP)	Epoxy resins
Polyvinyl Acetate (PVA)	Melamine resins
Polyvinyl Chloride (PVC)	Polyesters
Polystyrene (PS)	Polyurethane
Low density polyethelene (LDPE)	Urea – Formaldehyde
High density polyethylene (HDPE)	Alkyd resins

In this paper, the review of PET as an additive to the bituminous mixture in road making was presented in detail.

2. Polyethylene Terephthalate (PET)

PET is the most used thermoplastic polyester. PET is an acronym for polyethylene terephthalate, which is a long-chain polymer belonging to the generic group of polyesters. Polyethylene terephthalate (PET) is a semi-crystalline, thermoplastic polyester [6]. PET is one of the polyesters which formed by a polymerization reaction between an acid and alcohol [7]. PET is a polymer which easy to handle and also durable and strong, has low gas permeability, thermally stable and chemically [8]. With it good properties, PET was used widely in the form of the automobile part, lighting product, food packaging, electronics, sports tools, x-ray sheets, house ware, textile, power tools and photographic applications [7]. There are 60% of PET productions in term of bottles synthetic fibers [9].

3. Sources Of PET Wastes

Waste PET source can be subdivided into three which are foils, bottle, and cord from tires. Foils have two small problems with material recycling which is related to utilization of additive in production and molecular weight of PET. The bottle also has the same problem with foils, and it also has another one problem which is impurities problem. The cord from tires has big material recycling problem which is pollution of ground tire rubber and metals. Most of this waste is used as alternative fuel. Figure 1 shows the recycled textile fibre from used tired and Table 3 shows the compositions of waste tire cord.

Figure 1. Recycled textile fibre from used tires. [10]

PET can be used as an additive to the bituminous mixture in road construction which acts as the modifier for asphalt, and this can solve the waste PET recycling problem [11]. This initiative can increase lifespan and improve the engineering properties of modified materials.

Table 3. Compositions Of Waste Tire Cord [10]

Compositions	Weight percent (%)
Polyethylene therephthalate (PET)	77.6
Polyamide (PA)	18.7
Polypropylene (PP)	3.7

4. Utilization of Polymers In Asphalt Pavement

Asphalt has been used in road construction as the aggregate binder [12]. However, it is well known that asphalt mixture has several temperature weaknesses which are low temperature cracking damage, medium temperature fatigue and high-temperature rutting (Figure 2).

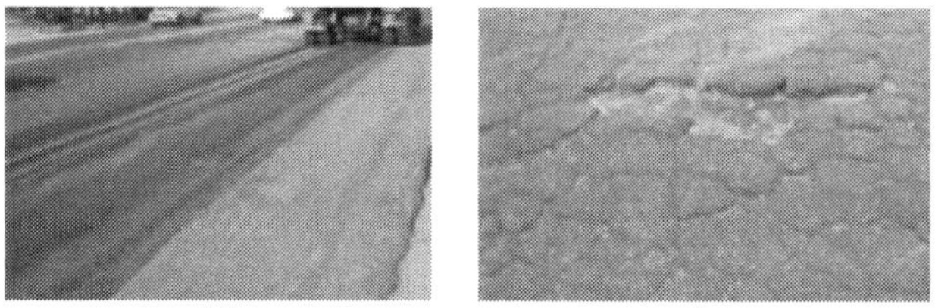

(a) (b)

Figure 2. (a) Rutting; (b) Fatigue cracking

Properties of asphalt can be improved by controlling the refines process or selecting the proper starting crude oil, which this two are very difficult to achieve [11]. Therefore, the most favoured method to improve the quality of asphalt is its modification. Asphalt becomes harder by air blowing. Asphalt can be softened by using diluents oils or fluxing agents. The other method to improve asphalt quality is the addition of polymers [11]. Rheological properties of asphalt are commonly improved by asphalt modification with a polymer [13]. Asphalt mixture should have a good property to resist rutting and shove under heavy load traffic and increasing of road users. Therefore, the stability of asphalt mixture is supposed to be good enough to handle traffic problem. The flow of the road surface depended on stability in the bituminous mixture. Flow is the capability of pavement bituminous mixture to adjust movement is subgrade and gradual settlement without cracking [14]. Asphalt viscoelastic properties depended on the chemical composition, and it will affect the performance of asphalt. Pavement defect did not only depend on traffic load but it also due to the ability of asphalt mixture to undergo temperature change which can cause cracking at low-temperature and rutting at a high-temperature region. Improvement of road performance is necessary because nowadays, there are increasing traffic factor, for example, higher traffic volume, higher tire pressure, and heavier loads. A better performance pavement needs asphalt which has good bonding between aggregate and less expose to low-temperature cracking and high-temperature rutting. Utilization of waste plastic and virgin plastic such as Polyethylene terephthalate (PET), polypropylene (PP), low density polyethylene (LDPE), ethylene vinyl acetate (EVA), acrylonitrilebutadienne-styrene (ABS), high density polyethylene (HDPE), Polystyrene-butadiene-styrene (SBS), rubber latex, Chloroprene Rubber (CR), polyolefines, Polyethylene (PE) and polyvinyl chloride (PVC) in asphalt mixture can improve the properties such as softening point, resistance to rutting, durability, viscoelastic property and fatigue life [15-25].

4.1 PET In Road Asphalt Mixture

Polymer modified asphalt which used virgin polymer to improve asphalt characteristic was already successfully done for many years [14]. Recently, the recycled polymer such as waste PET was used in substitution of commercial virgin material. Besides that, waste tire also has been used which it give positive effect to the pavement [14]. Using of the polymer such as PET in asphalt pavement whether with hot mix technique or cold mix technique is intended to improve stripping, thermal cracking, temperature susceptibility, fatigue damage and rutting resistance. They are used when better durability and performance are needed. Much of the time, they are chosen to decrease life cycle costs [11]. The polymer which used as modified binders gives a better improvement to cohesion and adhesion properties [15]. Action Program (WRAP) has mentioned that PET is the most plastic waste that can be recycled from a household waste stream [26]. It can be used as an additive to the bituminous mixture. There are many researchers that studies on this topic and many reports were published in the last years. A. Hassan, H. Ganjidoust, and Maghanaki [27] have used waste PET in the form of 3mm granules pallet in their studies. The waste PET was added as a partial fine aggregate with the bituminous mixture which used 60/70 bitumen grade and 12.5 mm aggregate grading. It was blended at 140°C-180°C and then compacted with 50 blows each side by using Marshall hammer. From these studies, the result proved that utilization of PET granulated (5% total weight of the bituminous mixture) to replace 20% of fine aggregate can improve the Marshall with the highest stability and lowest flow. Other than that, PET properties which are semi-crystalline and low glass transition temperature made it most effective as an additive [14].

E Ahmadinia and friends [17] conducts the evaluation on the effect of using waste PET to the engineering properties of SMA mixture. They carried out laboratory test on mechanical properties of asphalt mixture when 0%, 2%, 4%, 6%, 8%, and 10% of PET was blended into the mixture. The best proportion of PET that they have been found was 6% by weight of asphalt. Furthermore, utilization of PET gives a better effect on the properties of SMA, for example, increasing of stiffness and viscosity of asphalt at normal temperature. However, usage of PET as an additive was not given any satisfaction to improve the elasticity of asphalt significantly and on heating. There are three different forms of plastic waste have been used in road pavements which are in the forms of binder modifier, mixture reinforcement, and aggregate replacement. T. B Moghaddam and M. R. Karim [28] chose the second option for their studies which they have investigated the effect of adding waste PET into asphalt mixture. The results of the investigation showed that stability of asphalt mixture become higher when PET was added compared to the virgin mixture. The best PET amount is 0.4% to achieve better stability. However, the flow of the mixture would be increase when added a high amount of PET. Other than that, adding a higher amount of PET also made the stiffness of the mixture decreased which can cause increasing fatigue life. Z. N. Kalantar and his friends [15] have studied about utilization of PET as an additive into the bituminous mixture. The binder was blended at 150°C with five different proportion of PET which are 2%, 4%, 6%, 8%, and 10% by weight of optimum bitumen content. The results showed better resistance against rutting and permanent deformations after comparing it with the conventional binder. The results also showed that the increasing amount of PET would increase the softening point of the mixture. From this results, it can be concluded that tendency of binder to soften in hot weather would reduce if the resistance of the binder to the effect of heat increase. Although a lot effort have been made to test different type of plastic waste in asphalt mixture, however all those plastics cannot match the properties given by PET. There is no gas evolution when PET is heated in the temperature range of 120°C-165°C and it start decompose at temperature 270°C but there is still no harmful gas evolution [29].

4.2 PE In Road Asphalt Mixture

Awwad and Shbeeb on 2007 which used PE in their investigation and they are studies about the amount and best type of PE that should be used in the asphalt mixture [30]. There are two types of PE which are LDPE and HDPE and these used to coat the aggregate. From their results, it shows that grinded HDPE gives a better engineering property to the mixture. 12% by weight of asphalt content is the proposed amount of modifier. Added of this modifier also have increased the voids of mineral aggregate and air voids, reduce the density and increase the stability. C. Fuentes-Audén and friends [31] also studies the impact of PE on modified asphalt properties. From their research, the amount of PE used should not more than 5% because it will make viscosity which decreases the possibility of the mixture. Besides that, using of recycled PE can improve resistance to thermal fatigue, rutting and cracking. Donnchadh Caseya et al., 2008 [18] has studied about utilization of waste HDPE, LDPE, PET, ABS, and PVC in a bituminous mixture and they have concluded that LDPE and HDPE were the best recycled material to use. They have pointed out 4% is the ideal amount of recycled LDPE and HDPE as an additive to asphalt binder.

4.3 Fibre In Road Asphalt Mixture

Several investigations were made on the impact of fibre as an additive to the bituminous mixture. In Stone Mastic Asphalt (SMA), hot aggregate mixed is coated by hot asphalt then laid and rolled. Asphalt function as a binder. SMA has their disadvantages which can cause defective spots on the pavement where the water cannot flow over the roads. The cost of road construction will increase because of the process in using anti-stripping agents and it also has a limited use only [32-35].

Because of their great improvement effect, fibre becomes well known as a modifier. There are two types of common fibres modifier which are mineral fibres and cellulose fibres. They have been broadly utilized in various type of asphalt mixture, particularly in HMA with Open Grade Friction Concrete (OGFC) which also known as stone mastic asphalt (SMA) [36]. SMA has been used for over 20 years among European countries and if was created in the 1960s in Germany [37] & [38]. In 2004, B. J. Putman and Amirkhanian [39] did the comparison between waste CR from used tires and carpet fibres to the performance of SMA mixture. From their investigation, the result shows that the toughness of the mixture has been improved, but the moisture susceptibility or pavement deformation has no significant difference.

Shiuh and Kuie-Yi [40] was reported that polyester fibres are often used to increase durable and strength of asphalt fibre mastic at high temperature. They have investigated the performance of engineering properties of asphalt when used cellulose fibre and polyester as an additive. From their studies, the results show that the load-carrying ability of asphalt-fibre mastics can be increased when there is a good adhesion between asphalt and polyester fibres.

5. Conclusions

The major points that can be drawn from the data gathered from some scientific works of literature are as follows:

Based on scientific reports, experimental works and researches proved that plastic waste such as recycled PET suitable to be used to the bituminous mixture in road making. Utilizing of PET would improve properties of flexible pavement such as increasing of stability, stiffness and viscosity, hence it can improve stripping, thermal cracking, temperature susceptibility, fatigue damage and rutting resistance.

Last but not least, utilizing polymers as a modifier to the bituminous mixture would solve the problem of safe disposal of waste plastic materials in an eco-friendly way.

References

[1] *Plastic Common Wastes & Materials US EPA*. 2013; Available from: www.epa.gov/osw/conserve/material/plastic.htm.

[2] Plastics, E., *Facts 2014/2015: An analysis of European latest plastics production, demand and waste data*. 2013.

[3] Ecosystem, G., *A Study On Plastic Management In Peninsular Malaysia*. 2011, National Solid Waste Management Department Ministry Of Housing And Local Government Malaysia.

[4] S. T. Shy, W. Dipl, and F. Martin, *Current Practice of Municipal Solid Waste Management in Malaysia and the Potential for Waste-to-Energy Implementation*. 2009, Institute of Resource and Energy Technology, Technische Universität München, Germany.

[5] Amit Gawandea, G.Z., V.C.Rengea, Saurabh Taydea, G.Bharsakale, *An Overview On Waste Plastic Utilization In Asphalting Of Roads*. Journal of Engineering Research and Studies, 2012. **3**(2): p. 1-5.

[6] Webb, H.K., et al., *Plastic Degradation and Its Environmental Implications with Special Reference to Poly(ethylene terephthalate)*. Polymers, 2013. **5**(1): p. 1-18.

[7] Sinha, V., M.R. Patel, and J.V. Patel, *Pet Waste Management by Chemical Recycling: A Review*. Journal of Polymers and the Environment, 2010. **18**(1): p. 8-25.

[8] Awaja, et al., *Recycling of PET*. European Polymer Journal, 2005. **41**(7): p. 1453-1477.

[9] Ludwig Bottenbruch, S.A., *Engineering Thermoplastics: Polycarbonates, Polyacetals Polyesters, and Cellulose Esters*. 1996: Hanser Publishers.

[10] M. Sulyman, J. Haponiuk, and K. Formela, *Utilization of Recycled Polyethylene Terephthalate (PET) in Engineering Materials: A Review*. International Journal of Environmental Science and Development, 2016. **7**(2).

[11] B. Yuonne and M. P. M. R. Yajaira, *Polymer modified asphalt*. Vision Technologica, 2001. **9**(1): p. 39–48.

[12] Akmal, N. and A.M. Usmani, *Application Of Asphalt-Containing Materials*. Polym News, 1999. **24**: p. 136–40.

[13] Habib, N.Z., et al., *Rheological properties of polyethylene and polypropylene modified bitumen*. World Academy of Science, Engineering and Technology, 2010. **72**(12): p. 293-297.

[14] Kuloglu, N., *Effect of astragalus on characteristics of asphalt concrete*. Journal of Materials in Civil Engineering, 1999. **11**(4): p. 283-286.

[15] Z. N. Kalantar, M. R. Karim, and A. Mahrez, *A review of using waste and virgin polymer in pavement*. 2012, Construction and Building Materials. p. 55-62.

[16] R. Vasudevan, et al., *Utilization of waste polymers for flexible pavement and easy disposal of waste polymers*. International Journal of Pavement Research and Technology, 2010. **3**(1): p. 34-42.

[17] E Ahmadinia, et al., *Performance evaluation of utilization of waste Polyethylene Terephthalate (PET) in stone mastic asphalt*. Construction and Building Materials, 2012. **36**: p. 984-989.

[18] Donnchadh Caseya, et al., *Development of a recycled polymer modified binder for use in stone mastic asphalt*. Resources, Conservation and Recycling, 2008. **52**(10): p. 1167–1174.

[19] M. Garcia-Morales, et al., *Effect of waste polymer addition on the rheology of modified bitumen*. Fuel, 2006. **85**(7-8): p. 936-943.

[20] Sinan Hınıslıoğlu and E. Ağa, *Use of waste high density polyethylene as bitumen modifier in asphalt concrete mix.* Materials Letters, 2004. **58**(3–4): p. 267–271.

[21] Salter, R.J. and F. Rafati-Afshar, *Effect of additives on bituminous highway pavement materials evaluated by the indirect tensile test*, in *66th Annual Meeting of the Transportation Board.* 1987, Transportation Research Board: Washington District of Columbia, United States. p. 183-195.

[22] S. Bose and P.K. Jain., *Laboratory studies on the use of organic polymers in improvement of bituminous road surfacing*, in *Highway Research Bulletin*. 1989, Indian Roads Congress: New Delhi. p. 63-79.

[23] Little, D.N., *An Additive Of Asphalt Additives To Reduce Permanent Deformation And Cracking In Asphalt Pavements: A Brief Synopsis Of Ongoing Research.* Proc. the Association of Asphalt Paving Technologists, 1986. **55**: p. 314-320.

[24] T. Süreyya, O. Halit, and A. Atakan, *Investigation of rutting performance of asphalt mixtures containing polymer modifiers.* Constr Build Mater, 2007. **21**(2): p. 328–37.

[25] G N King, H W Muncy, and J B Prudhomme, *Polymer modification: Binder's effect on mix properties*, in *Association of Asphalt Paving Technologists Proc*, E.L.S. Jr, Editor. 1986. p. 519-540.

[26] WRAP, W.a.R.A.P., *Annual local authorities plastics collection survey*. 2007.

[27] A. Hassan, H. Ganjidoust, and A.A. Maghanaki, *Use of plastic waste (polyethylene terephthalate) in asphalt concrete mixture as aggregates replacement.* J. Waste Managements and Research, 2005. **23**(4): p. 322–327.

[28] Baghaee Moghaddam, T., M.R. Karim, and T. Syammaun, *Dynamic properties of stone mastic asphalt mixtures containing waste plastic bottles.* Construction and Building Materials, 2012. **34**: p. 236-242.

[29] Menaria, Y. and R. Sankhal, *Use Of Waste Plastic In Flexible Pavement-Green Roads.* Open Journal of Civil Engineering, 2015. **5**: p. 299-311.

[30] Awwad, M.T. and L. Shbeeb, *The use of polyethylene in hot asphalt mixtures.* American Journal of Applied Sciences, 2007. **4**(6): p. 390-396.

[31] C. Fuentes-Audén, et al., *Evaluation of thermal and mechanical properties of recycled polyethylene modified bitumen.* Polymer Testing, 2008. **27**(8): p. 1005–1012.

[32] T. S. Shuler, J. H. Collins, and J.P. Kirkpoatrick., *Polymer Modified Asphalt Properties Related To Asphalt Concrete Performance.* Asphalt Rheology Relationship to Mixture. 1987: Philadelphia.

[33] N. F. Moghadas, E. Aflaki, and M.A. Mohammadi, *Fatigue behavior of SMA and HMA mixtures.* Construct Build Mater, 2010. **24**: p. 1158–1165.

[34] S. E. Zoorob and L. B. Suparma, *Laboratory design and investigation of the properties of continuously graded Asphaltic concrete containing recycled plastics aggregate replacement (Plastiphalt).* Cement and Concrete Composites, 2000. **22**(4): p. 233-242.

[35] Denning, J.H. and J. Carswell, *Improvements in rolled asphalt surfacing by the addition of organic polymers*. 1981: Transport and Road Research Laboratory.

[36] J. P. Serfass and J. Samanous., *Fiber-Modified Asphalt Concrete Characteristics, Application And Behavior.* Journal of Association of Asphalt Paving Technologists, 1996. **65**: p. 193-230.

[37] M. I. Asi, *Laboratory Comparison Study For The Use Of Stone Matrix Asphalt In Hot Weather Climates.* J Constr Build Mater, 2006. **20**: p. 982–989.

[38] D. L. Bischoff and R. B. Schmiedlin, *Stone matrix asphalt: The wisconsin experience.* Winsconsin, Department Of Transportation, 2002(January): p. 25.

[39] B. J. Putman and S.N. Amirkhanian. *Utilization of waste fibers in stone matrix asphalt mixtures*. in *Resources, Conservation and Recycling*. 2004.

[40] Shiuh, J. and L. Kuie-Yi, *Mechanisms And Behaviour Of Bitumen Strength Reinforcement Using Fibers.* J. Mater. Sci, 2005. **40**: p. 87-95.

IOP Conf. Series: Materials Science and Engineering **203** (2017) 012005 doi:10.1088/1757-899X/203/1/012005

Utilization of polyethylene terephthalate (PET) in bituminous mixture for improved performance of roads

A. F. Ahmad[1], A. R. Razali[1], I. S. M. Razelan[2], S.S. A. Jalil[2], M.S. M. Noh[2], A. A. Idris[2]

[1]Faculty of Mechanical Engineering, Universiti Malaysia Pahang, 26600 Pekan, Pahang, Malaysia
[2]Faculty of Civil Engineering a and Earth Resources, University Malaysia Pahang, Lebuhraya Tun Razak, 26300 Gambang, Kuantan, Pahang

E-mail: amaninafarhana.ahmad@gmail.com

Abstract. Plastic bottle for recycling can be found from the household waste stream, and most of them are made from Polyethylene Terephthalate. In this research, PET is utilized to explore the potential prospects to upgrade asphalt mixture properties. The objectives include deciding the best measure of PET to be used. For experimental, Marshall mix design was utilized to determine the ideal bitumen binder content and to test the modified mixture properties. The samples were created per the requirement for aggregate course wearing (ACW14) using the Standard Specification of Road Work (SSRW) in Malaysia. 20 samples were utilized to determine the binder content, and 30 samples were used to research the impact of modifying asphalt mixtures. 2%, 5%, 10%, 15% and 20% of PET by weight of the optimum binder content (4.8%) were tested. Optimum PET content is 10%, and the result shows a good stability with 16.824kN, 2.32g/cm3 bulk density, void filled with bitumen (VFB) with 71.35%, flow with 3.2248mm, air void (AV) with 4.53%, and void of mineral aggregate (VMA) with 15.15%. The outcomes showed that PET modifier gives better engineering properties. Therefore, 10% of PET by the weight of binder content was suggested as the best amount of the modifier.

1. Introduction

As urbanization keeps its pace with a steady growth, waste management has turned into a noteworthy environmental issue around the world. Indeed, Malaysia is one of the biggest plastics makers in Asia, with more than 1,550 manufacturers, utilizes around 99,100 people, [1] is also facing the issue. Fast development of plastics industries in Asia's emerging markets particularly in Malaysia, in one way it raised expectations for everyday comforts, in the other way it also causes an environmental issue. This issue tends the government to bans on bags and another packaging. Restrictions of plastic packaging are being enacted or proposed over the continent, including in Philippines, India, and Malaysia. Due to the manufacturing of plastic which causes chemical plant pollution, worldwide assertive citizens protested its presence [2]. It is identified that plastic waste contamination in China alone is to hit 19.1million tons by 2017 [3]. This projection can also be used to predict the number of amounts of plastic waste in Malaysia for the same year. Recycling this plastic into another form might be one of the cheapest and rapid solutions.

Most of the road networks in Malaysia are flexible pavement design. The number of components (binder, aggregate and additive) in the mixture and the properties of it can influence the performance of flexible course. Bitumen can be altered by adding distinct sorts of additive [4]. Polymers are one of the additives which mainly having good attention to be used on earthwork and building construction field. PET is one of the polymers that vastly used to coat aggregates in asphalt mixtures. Coating ensures a proper bond between aggregates and improving its surface roughness for a good relationship thus makes superior engineering properties to asphalt mixtures [5, 6].

In this research, PET which is a type of plastomer was used to modify the properties of the asphalt mixture. It is used to coat the aggregate instead of changing the properties of bituminous. The primary objectives of this research are to study the effect on the hot mix asphalt when PET was added and to determine the optimum bitumen content and excellent PET content in the asphalt mixture.

2. Materials and method

Bituminous material: Asphalt binder 80/100 was used as the main mixture medium. Some laboratory tests have been carried out to determine the properties of the bitumen: Thin Film Oven, Penetration, Ductility, and Softening point. Table 1 shown the properties of asphalt binder, which are in a particular range of asphalt grade 80/100.

Aggregate properties: The laboratory tests performed on aggregates were: Los Angeles Abrasion, Aggregate Impact Value, and Aggregate Crushed Value. Table 2 shown the results of aggregate tests. Standard Specification of Road Work in Malaysia [7] used to define the mix gradation limits as shown in Table 3.

Polyethylene Terephthalate (PET): PET is an acronym for polyethylene terephthalate, which is a long-chain polymer belonging to the generic group of polyesters. Polyethylene Terephthalate (PET) is a semi-crystalline, thermoplastic polyester [8]. PET is one of the polyesters which formed by a polymerization reaction between an acid and alcohol [9]. PET is a polymer which easy to handle and also durable and sturdy, has low gas permeability, thermally stable and chemically [10]. For this research, plastic bottle which made by PET was used. The bottles were collected, and shred into small pieces about 2 mm to 5mm. Table 4 shows mechanical properties of PET.

Table 1. Properties of used asphalt binder

Test	Test Result	Specification	
		Min	Max
Ductility (cm)	150	100	-
Penetration (0.1mm)	86	80	100
Softening point (°C)	46.3	45	52
Thin Film Oven (% wt)	0.398	-	0.75

Table 2. Properties of used Aggregate

Test	Average % Loss
Los Angeles Abrasion	21.24
Aggregate Impact Value	21.55
10% Fine Aggregate	11.60
Aggregate Crushed Value	25.46

Table 3. Proposed mix gradation

% Passing	20 mm	14 mm	10 mm	5 mm	3.35 mm	1.18 mm	0.425 mm	0.15 mm	0.075 mm
Proposed Mix	100	95	81	56	47	26	18	10	6
Minimum Limit	100	90	76	50	40	18	12	6	4
Maximum Limit	100	100	86	62	54	34	24	14	8

Table 4. Mechanical Properties of PET [10]

Properties	**Values**
Average molecular weight	30000 - 80000 $gmol^{-1}$
Density	1.41 gcm^{-3}
Melting temperature	255 - 265 °C
Glass transition temperature	69 - 155 °C
Young's modulus	1700 MPa
Water absorption (24 h)	0.5%

After determining the optimum bituminous material content (4.8% of mixture weight), five different proportion of PET (2%, 5%, 10%, 15% and 20% by weight of binder content) has been added to the asphalt mix. There are five samples for every PET portion.

Sample preparation: Determination of the right proportion of aggregate and asphalt can affect the performance of an asphalt mixture. 20 samples with 1116 grams each was prepared per the proposed mix design to determine the maximum bituminous content that would produce asphalt mixtures with strength and durability properties that meet the Standard Specification of Road Work in Malaysia. Three sample of each binder content proportion were used to prepared asphalt mix. After that, the average of unit weight, Marshall stability and flow properties for each binder content were determined. Seven binder contents were considered (4%, 4.5%, 5.0%, 5.5%, and 6.0%). The ACW14 suggests that the optimum bitumen content ought to obtain in a range of 3 to 5% air void, 70% to 80% void filled with bitumen and higher than 8000N of stability. From the experiment, the optimum bituminous content was 4.8%.

25 samples of asphalt mixtures were prepared with this binder content to test the effect of adding the PET to the mix. There are five proportions of the PET (2, 5, 10, 15, and 20%) which were added to these 25 samples. The samples were tested to determine the unit weight, stability and flow for each PET proportion. The aggregate of each specimen is heated until it reached 170° C before adding the PET. The duration and heating temperature were chosen based on literature review from reference [11]. This temperature is sufficiently hot to melt the PET with a particular size of 2 mm - 10 mm to such an extent that it would stick to the aggregate surfaces and leave a textured PET surface with a bond between the coated aggregates. The overall binder is assumed as 4.8% which same as optimum bitumen content. The weight of the overall binder is calculated as Eq. (1):

$$\%binder = \frac{Wb}{Wa + Wb} \tag{1}$$

Where

Wb is weight of binder
Wa is weight of aggregate

By using the formula, the weight of overall binder is 56.27g. The amount of bitumen decreases when the amount of PET increases, which to achieve 4.8% as the overall binder. Table 5 shows the proportion of the mix.

Table 5. Proportion of the modified asphalt mixture

Binder				
plastic		Bitumen		overall
%	g	%	g	g
2	1.13g	98	55.14g	
5	2.81g	95	53.46g	
10	5.623g	90	50.64g	56.27
15	8.44g	85	47.83g	
20	11.25g	80	45.02g	

3. Results

A comparison between conventional asphalt mixture (No PET) and PET asphalt mixture is presented below.

3.1. Bulk density- PET content relationships

The bulk density of the conventional asphalt mixture is higher than modified asphalt mixture. The bulk density of modified asphalt mixture reaches a maximum value when PET content is around 2%. The general trend shows that as the content of PET increased, the bulk density also increased. Figure 1 shows the bulk density of PET modified asphalt.

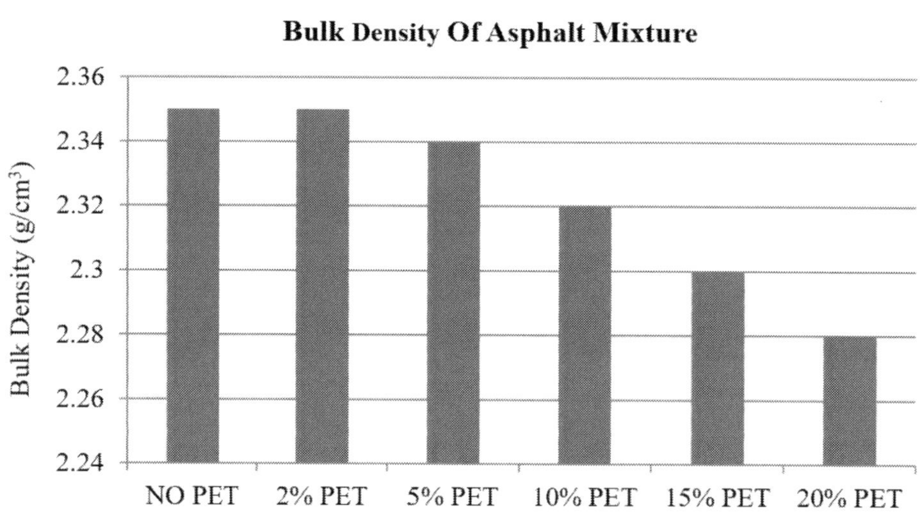

Figure 1. Bulk Density of PET Modified Asphalt

3.2. Stability- PET content relationship

The stability of the conventional asphalt mixture is lower than modified asphalt mixture. 18.359kN is the highest balance for asphalt mixture which was treated with 20% of PET. Figure 2 show the stability of modified asphalt mixture increase continuously by an increase of the modified content.

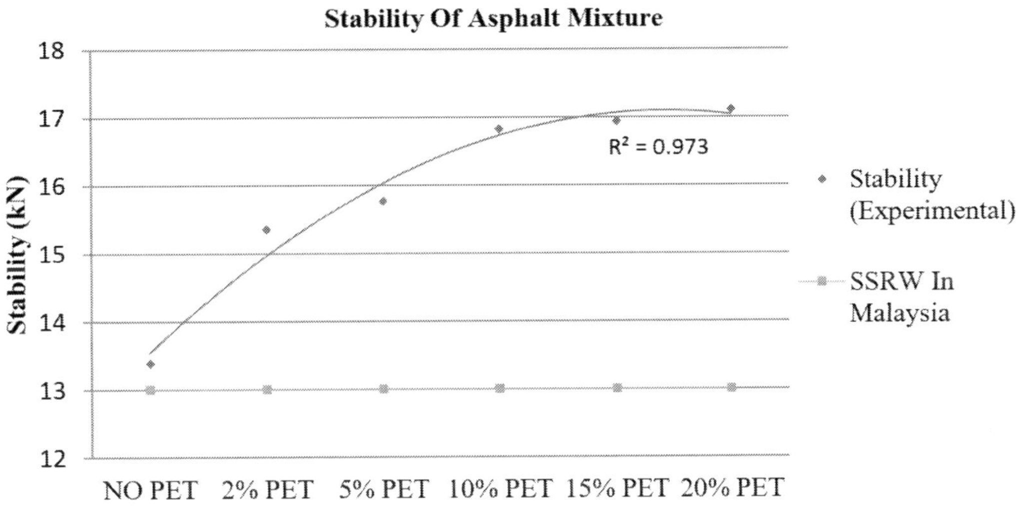

Figure 2. Stability of modified asphalt mixture

3.3. Flow- PET content relationship

The flow of the conventional asphalt mixture is lower than modified asphalt mixture. 3.374mm is the highest flow can be observed which treated with 20% PET. This value is still within the range of Standard Specification of Road Work in Malaysia. Figure 3 show that when the % of PET increased, the flow increased continuously.

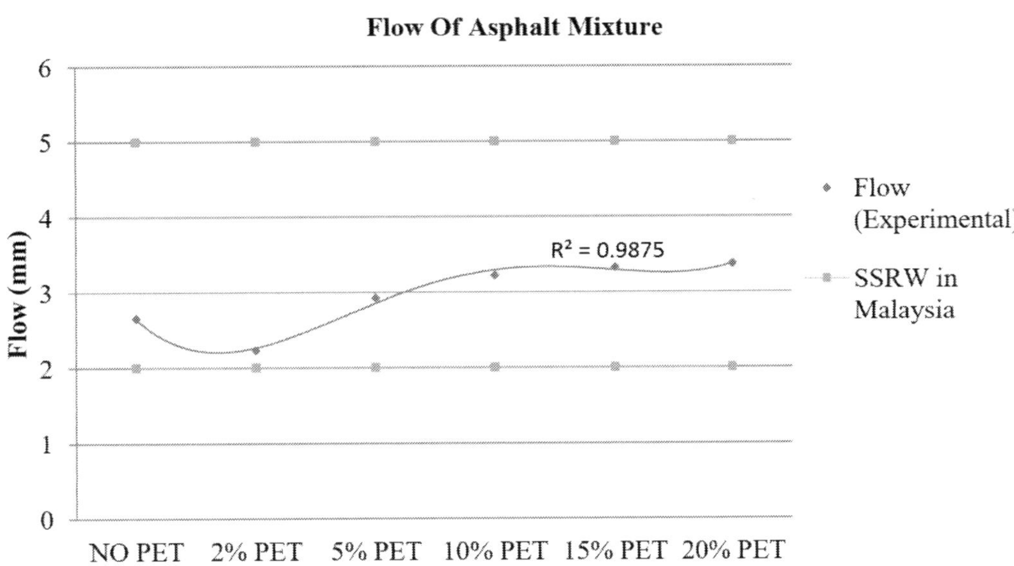

Figure 3. Flow of modified asphalt mixture

3.4. Air Void (AV) - PET content relationship

Air void is also known as void in the total mixture. From the experiment, the air void proportion of the conventional asphalt mixture is lower than modified asphalt mixture. The result of air void with PET content 15% and 20% exceeded the range of Standard Specification of Road Work in Malaysia. The only mixture with 2%, 5% and 10% of PET have achieved the specification. When the modifier content increased, it was found that the amount of air void was also increased (Figure 4).

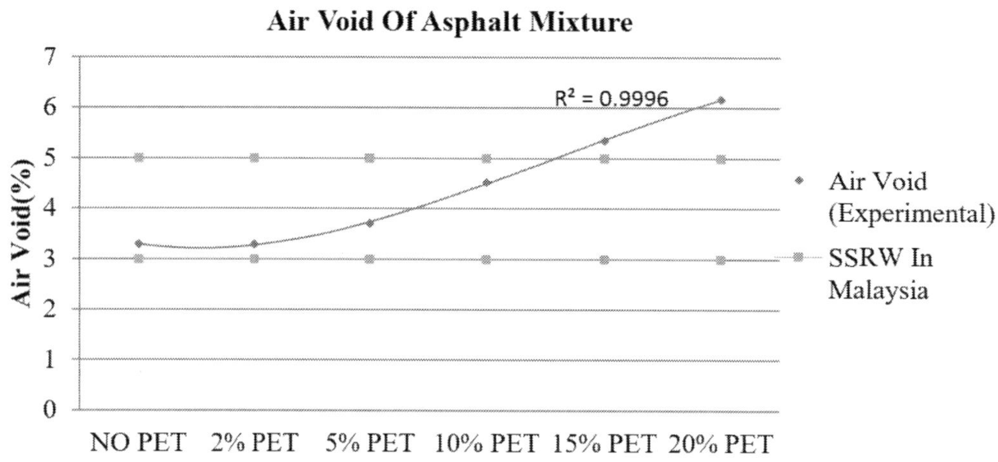

Figure 4. Air Void of modified asphalt mixture

3.5. Void filled with bitumen(VFB) - PET content relationship

In general, the VFB of conventional asphalt mixture is higher than a mixture with PET content. VFB of 15% and 20% PET content exceeded the Standard Specification of Road Work in Malaysia. The only mixture with 2%,5%,and 10% of PET has achieved the specification. The content of VFB in modified asphalt mixture decreased when the amount of PET increased (Figure 5).

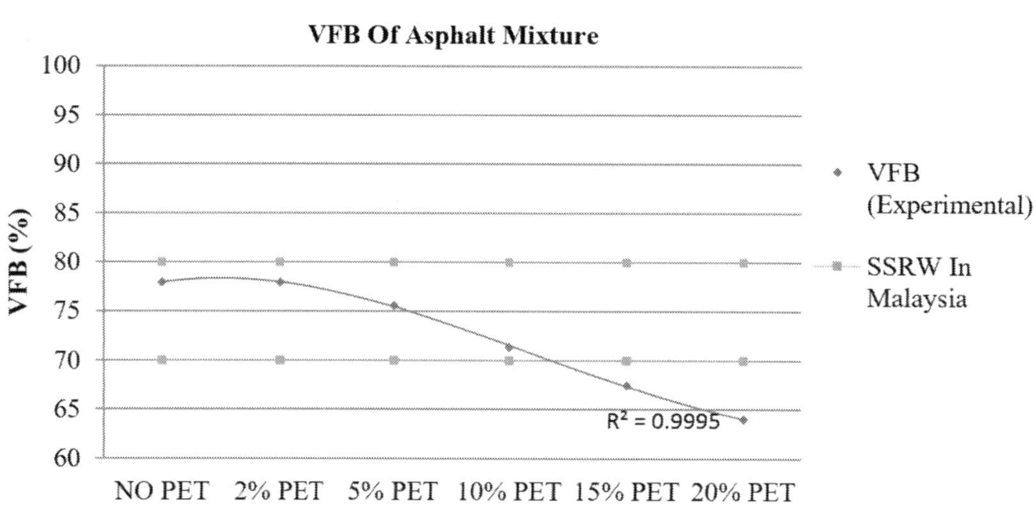

Figure 5. The amount of VFB in modified asphalt mixture

3.6. Void of mineral aggregate (VMA) - PET content relationship

VMA of conventional asphalt mixture is lower than modified asphalt mixture. The maximum of VMA is 16.61% at 20% of PET content. The trend shows that the VMA increased as the percent of PET increased. Figure 6 show the trend of VMA in modified asphalt content.

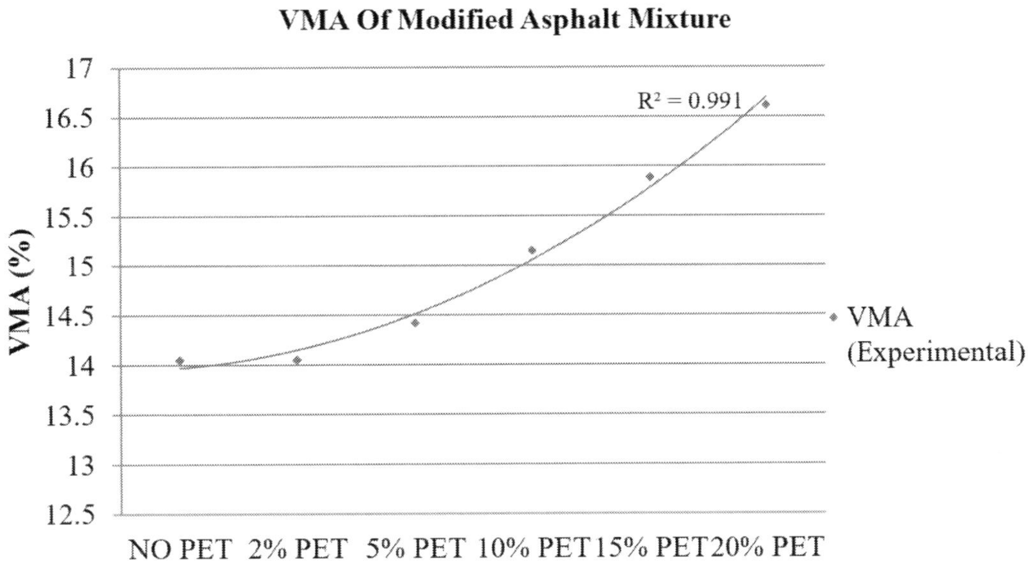

Figure 6. VMA in modified asphalt mixture.

3.7. Optimum Modifier Content

Based on research done by [12] asphalt mix with optimum modifier content should have maximum stability, maximum bulk density and percent of air voids within the allowed range of specifications. The procedures to obtain optimum modifier content is by plot the graph of Marshall properties which are graph stability versus modifier content, graph density versus modifier content, and graph voids in the total mix (VTM) versus modifier content. From the graphs, the optimum modifier content has been determined by finding the modifier content at maximum stability, maximum density, and Air Voids within the allowed range of specifications. The optimum modifier content was calculated as Eq (2) [12] and 4% of VTM is chosen because it is median of air voids range. The optimum modifier content is selected as the content that satisfies the Standard Specification of Road Work in Malaysia (table 6).

$$OMC = \frac{(MC \max density) + (MC \max stability) + (MC \ at\ 4\% \ of\ VTM)}{3} \quad (2)$$

Where

OMC is optimum modifier content
MC is modifier content
VTM is air voids of total mix

Table 6. Standard Specification of Road Work in Malaysia for ACW14

Parameter	Wearing Course
Stability, S	>13000N
Flow, F	2-5 mm
Stiffness, S/f	> 2600N/mm
Air voids in mix (VTM)	3-5%
Voids in aggregate filled with bitumen (VFB)	70-80%

From the calculation result, optimum modifier content was decided with 10% PET content. Since10% PET content has satisfied the Standard Specification of Road Work in Malaysia.

4. Discussions

4.1. Stability

Based on the results of the study, the stability of modified mixture is higher compared to the non-modified mixture. This matter would positively have an impact on resistance and rutting of this mixture. When waste plastic is used in coating aggregate, it improved the stability of the asphalt mixture [13]. Besides that, utilization of plastic in the dry process (aggregate coating) can also minimize the susceptible rutting compared to conventional asphalt mixture. Rutting occurs due to weak shear strength to withstand continual heavy load and affect the performance in the pavement surface course. This problem makes the mixture accumulate and forming a rut characterized by a downward and lateral movement of the pavement [14]. By adding PET to the bituminous mixture, shear strength and stability will increase, and this can overcome the rutting problem.

Studies made by Mishra, B. and R.S. Mishra [15] show that waste plastic that added directly with hot aggregate can increase the stability of asphalt mixture and this result also achieve by Sridhar, Bose, Kumar, and Sharma on 2004 [16]. Vasudevan and his friends [6] also stated that this method was better compared to the wet process. In the wet process, polymer particles are added to the bitumen at a certain temperature to produce polymer modified binder. In the dry process, the polymer particles are added directly to coated aggregate before the binder is added to the mixture [17]. The bituminous mixture produced by the dry process has shown that the polymer does not swell during mixing period, transportation and laying. The rate of bitumen absorption by polymer is high which can make the residual bitumen elastic and stiffer, and this will change the rigidity of the polymer whereby effect the performance of the pavement [18].

4.2. Air Void

The proportion of air void can influence the performance of the asphalt mixture. This is because high air void can cause sensitive crack pavement because of asphalt oxidation or the water enter the voids which will cause striping of aggregate [19]. Air voids are pockets of air or little air spaces which can be found between the coated aggregate particles in the last compacted mixture. A particular amount of air void is needed in all dense-graded highway mixes to consider some additional pavement compaction under traffic activity and to give spaces into which little amount of asphalt can stream during this subsequent compaction [20]. Air void exceedingly related to the compaction to the mixture. Air void decreased when the small particles were filled with aggregate, and this happens when high compaction is applied. When shredded PET was added into the hot aggregate, it might not be entirely melted, and compaction may cause void whereby the aggregate is not filled by the particle properly.

PET will be not fully melted when the PET content was increased. Hence the aggregates cannot be coated properly. Besides that, the bitumen content was reduced as the PET content increased to achieve the amount of 4.8% binder content. This affected the bonding between aggregate and caused increasing of the void. 3% to 5% of air void were required to permit the additional compaction occurred because of traffic load. Lower voids in the mixture can cause bleeding to occur because the pavement cannot withstand extra load reacted on the mix surface.

4.3. Bulk Density

Density is one of the most important parameters in construction of asphalt mixtures. A good mixture contains enough air voids to prevent rutting due to plastic flow but low enough air voids to prevent permeability of air and water. Since density of an asphalt mixture varies throughout its life the voids must be low enough initially to prevent permeability of air and water and high enough after a few years of traffic to prevent plastic flow. There are three primary methods of specifying density: percent of control strip, percent of laboratory density, and percent of theoretical maximum density. All three methods can be used to obtain satisfactory compaction if used correctly. Some effort construct a control test strip, measure the density on the strip, and use that density as the target density for the project [21]. Other efforts compact samples in the laboratory during mix design and during construction and use that density as the target density [22]. Finally, other efforts measure the theoretical maximum density (ASTM D 2041) and use some percentage of that density as the target density [23, 24] . The voids in an asphalt mixture are directly related to density; thus, density must be closely controlled to ensure that the voids stay within an acceptable range. High voids lead to permeability of water and air resulting in water damage, oxidation, gravelling, and cracking. Low voids lead to rutting and shoving of the asphalt mixture [25]. The acceptable range based on the Standard Specification of Road Work in Malaysia is between 3% to 5%.

5. Conclusions

The outcome from the lab testing shows that stability, flow, voids of mineral aggregates and air voids of modified asphalt mixture is higher than conventional asphalt mixture. The results also show that bulk density and void filled with bitumen of modified asphalt mixture is lower than conventional asphalt mixture. The optimum bitumen content is 4.8%, and excellent PET content is 10%. All the results uncover the modified asphalt mixture with 10% PET content is better than conventional asphalt mixture.

In term of economic value, utilization of recycled PET could reduce road construction cost because this material is not expensive compared to bitumen and it also easy to found. Therefore PET modified asphalt mixture can oppose the previously road failures. Additionally, it enhances the level of performance and the lifespan of the road. In conclusion, utilization of PET in asphalt mixture can give more superiority compared to asphalt mixture without PET which known as conventional asphalt mixture. Having considered the economic and environmental prudent angles, utilization of PET as an additive to asphalt mixture is suitable to be used for road pavement.

References

[1] MGCC, *Market Watch 2012 The Malaysian Plastic Industry*. 2012, Malaysian-German Chamber of Commerce and Industry (MGCC).

[2] Toloken, S. *waste And Pollution Concerns Top Asia Plastics Forum*. 2011; Available from: [http://www.plasticsnews.com/article/20110921/NEWS/309219983/waste-and-pollution-concerns-top-asia-plastics-forum].

[3] Zhang, H. and Z.G. Wen, *The consumption and recycling collection system of PET bottles: a case study of Beijing, China*. Waste Manag, 2014. **34**(6): p. 987-98.

[4] Bambang S Subagio, Djunaedi Kosasih, and Busnial, *Development of stiffness modulus and plastic deformation characteristics of porous asphalt mixture using tafpack super*. Proceed. Eastern Asia Soc.Transportation Studies, 2005. **5: 803 – 812**.

[5] Min-Gu Kim, Joe W. Button, and D.-W. Park, *Coatings to improve lowquality local aggregates for hot mix asphalt pavements*. Texas Transport. Inst., Texas A&M Univ. System., 1999. **SWUTC / 99 / 167405 - 1**.

[6] Vasudevan, R., et al., *Utilization of Waste Plastics in Construction of Flexible Pavements*. Indian Highways, 2006. **34**(7): p. 5-20.

[7] JKR, M., *Standard Specification For Road Work In Malaysia*, J.K.R. Malaysia, Editor. 2008.

[8] Webb, H.K., et al., *Plastic Degradation and Its Environmental Implications with Special Reference to Poly(ethylene terephthalate)*. Polymers, 2013. **5**(1): p. 1-18.

[9] Sinha, V., M.R. Patel, and J.V. Patel, *Pet Waste Management by Chemical Recycling: A Review*. Journal of Polymers and the Environment, 2010. **18**(1): p. 8-25.

[10] Awaja, et al., *Recycling of PET.* European Polymer Journal, 2005. **41**(7): p. 1453-1477.
[11] Menaria, Y. and R. Sankhal, *Use Of Waste Plastic In Flexible Pavement-Green Roads.* Open Journal of Civil Engineering, 2015. **5**: p. 299-311.
[12] El-Saikaly, M.A., *Study of the Possibility to Reuse Waste Plastic Bags as a Modifier for Asphalt Mixtures Properties (Binder Course Layer)*, in *Faculty of Engineering Civil Engineering / Infrastructure*. 2013, The Islamic University of Gaza.
[13] Chavan, M.A.J., *Use Of Plastic Waste In Flexible Pavements.* International journal of Application or Innovation in Engineering & Management (IJAIEM), 2013. **2**(4).
[14] Rahman, W.M.N.W.A. and A.F.A. Wahab, *Green Pavement Using Recycled Polyethylene Terephthalate (PET) as Partial Fine Aggregate Replacement in Modified Asphalt.* Malaysian Technical Universities Conference on Engineering & Technology 2012 (Mucet 2012), 2013. **53**(124-128): p. 124-128.
[15] Mishra, B. and R.S. Mishra, *Study on Use of Waste Plastic Materials in Flexible Pavements.* Research in Science, Engineering and Technology 2015. **4**(8).
[16] Sridhar, R., et al., *Performance Characteristics of Bituminous Mixes Modified by Waste Plastic Bags.* Highway Research Bulletin 2004: p. 1-10.
[17] Kim, Y.R., *Asphalt Pavements*. Vol. 1. 2014: CRC Press.
[18] Airey, G.D., M.M. Rahman, and A.C. Collop, *Effect of Rubber Bitumen Interaction on the Mechanical Durability of Impact Absorbing Asphalt.*
[19] International Journal of Pavement Engineering, 2003. **4**(2): p. 105-119.
[20] R.Hmoud and Haydar, *Evaluation of VMA and Film Thickness Requirements in Hot-Mix Asphalt.* Modern Applied Science, 2011. **5**(4): p. 166-176.
[21] Speight, J.G., *Asphalt Materials Science and Technology.* 2016: p. 437-474.
[22] Hoppe, J.E., et al., *Feasibility of Reclaimed Asphalt Pavement (RAP) Use As Road Base and Subbase Material.* A Report: Virginia Center for Transportation Innovation and Research University of Virginia, Charlottesville, Virginia VCTIR 15-R6, 2015: p. 1-42.
[23] Lo Presti, D., et al., *Laboratory Mix Design of Asphalt Mixture Containing Reclaimed Material.* Advances in Materials Science and Engineering, 2014. **2014**: p. 1-11.
[24] Nam Tran, P. Turner, and J. Shambley, *Enhanced Compaction To Improve Durability And Extend Pavement Service Life: A Literature Review*. 2016, National Center for Asphalt Technology Auburn University, Auburn, Alabama.
[25] Sayed, M.A.G.E., *Effect of Changing Theoretical Maximum Specific Gravity on Asphalt Mixture Design.* Engineering Journal 2012. **16**(4).
[26] Brown, E.R. and C. Steve. *A Study of In-Place Rutting of Asphalt Pavements*. in *Annual Meeting of the Association of Asphalt Paving Technologist*. 1989. Nashville.

IOP Conf. Series: Materials Science and Engineering **203** (2017) 012006 doi:10.1088/1757-899X/203/1/012006

Finite Element Structural Analysis of a Low Energy Micro Sheet Forming Machine Concept Design

A.R. Razali*, C.T. Ann, A.F. Ahmad, H.M. Shariff, N.I. Kasim, M.A. Musa

Faculty of Mechanical Engineering, Universiti Malaysia Pahang, 26600 Pekan, Pahang, Malaysia

*Email: akhtar@ump.edu.my

Abstract. It is forecasted that with the miniaturization of materials being processed, energy consumption will also be 'miniaturized' proportionally. The aim of this researchis to design a low energy micro-sheet-forming machine for the application of thin sheet metal. A fewconcept designsof machine structure were produced. With the help of FE software, the structure is then subjected to a forming force to observe deflection in the structure for the selection of the best and simplest design. Comparison studies between mild steel and aluminium alloys 6061 were made with a view to examine the most suitable material to be used. Based on the analysis, allowable maximum tolerance was set at 2.5μm and it was found that aluminium alloy 6061 suffice to be used.

1. Introduction

Rapid development of micro-manufacturing technologies for the manufacture of individual parts and systems was mainly driven by the increased demands on micro-products.Development works in the micro-manufacturing technologies includes development of new manufacturing processes, tools and machineries. Manufacturer nowadays still largely relies on conventional techniques based on materials removal, either by thermal, mechanical or by chemical means. To-date, tremendous efforts have been made to improve efficiency of micro-manufacturing technologies specifically in its processes [1-2]. The effortshowever, have not resulted radical change to the technologies but created a stigma of micro-manufacture as an 'expensive' and 'wasteful' business. Conversion of micro-materials into end products by high-rate plastic deformation (micro-forming) would address two key issues which are of importance for the industry, namely, reduction of manufacturing costs and improvement of product quality [3-4]. Besides, the need is proportional to the complexity of forming of micro-materials which requires good understanding of machines and processes at micro-scale.

A research group elsewhere [4]was making significant efforts in this field with a view to achieve the goals. A micro-sheet-forming machine has been developed by the group and proven to have capability of producing micro-products with consistent accuracy and precision [5]. The machine used linear motor to drive the forming ram/punch. It was reported that 5kN of stamping force was achieved. Forming capability was at 840 parts per minute. The machine required3 phase 415VAC electricity supply to work. The story had demonstrated success in micro-machines development,however,effort to tackle energy conservation was not addressed. It is expected that with the miniaturization of material being processed, energy consumption is also 'miniaturized' proportionally. Miniaturization of part with a scale of 100:1, could also translate the energy reduction within the same ratio [6].

A low power micro-sheet-forming machine which uses hundred times less energy than its conventional counterpart is practically a direct solution to address the issue. This is achieved by using a low power actuation device to perform the forming process. Solenoid is one of the promising solution to be used for the development work addressed here. With rated power of 12-24VDC, the solenoid is capable of producing kilonewton force over millimeters' distance. This setup could be a practical solution to reduce energy usage while performing sheet forming process.

2. Machine Design Specification/Consideration

The following factors were considered when designing the machine specifications: forming of thin sheet metal with thickness of less than 100μm; forming tolerance is 5~10 % of the strip thickness (i.e. 50μm thick strip was used); electricity power used is direct current approximately 12-24VDC; flexible design for easy set-up and changes of the forming tools, fixtures, part carrier and parts collection.

3. Machine System Development

3.1 Machine Frame

Various machine structural designs were examined. Thus, various machine concept layouts were created as depicted in Figure 1(a) to (d). Evaluation of the various machine concepts was carried out by reviewing deflection on each part of the frames, and then isfollowed by eliminating the weakest structure.The analysis was carried out by exerting200N of forceat four different points on top of various frame concepts with a view to simulate the forming force. O-frame machine design was selected as it produced the lowest structural deflection.

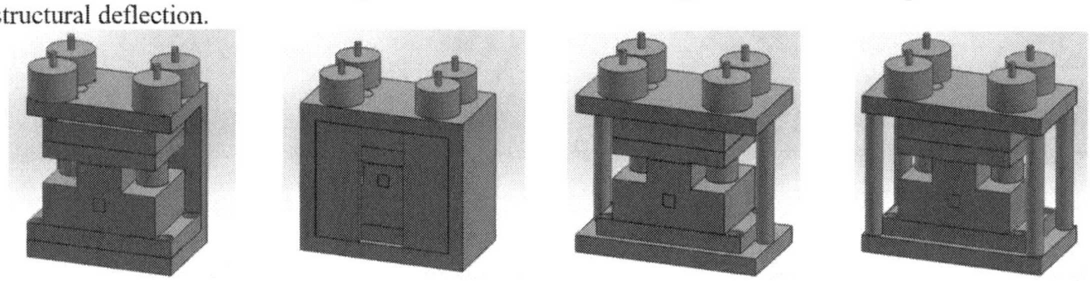

a) C-type b) O-type c) 2-column type d) 4-column type

Figure 1. Various machine frame design.

3.2 Tool Development

The tooling system basically consists of an upper and a lower half die with die-inserts (Figure 2a). The lower half die was fixed to the forming machine frame. Upper and lower half dies were linked together with two linear guide posts. The guide postswere connected to the upper die by push-fit assembly. Four springs were used to linkbetween blank and punch holder plates. The die insert located at the middle of punch guide plate and it is aligned with the punch with the use of the linear guide posts. A tunnel shaped drawer was made below the die-insert plate for parts collection.

3.3 Control System

Linear actuation was realized by using four linear solenoids as depicted in Figure 2(b). Each solenoid hasa capability of thrusting 170N of force at 25% duty cycle. The solenoids were powered by 24VDC power supply unit and controlled with a pair of solid state relays. Each relay connected to two solenoids and controlled by a microcontroller. The system receives 240 VAC as an incoming electricity supplyand filtered by an EMF filter. Filtered power source is fed to an energy meter to monitor energy usage by the machine. Figure2(b) shows the schematic illustration of connection of the main machine elements and system for control.

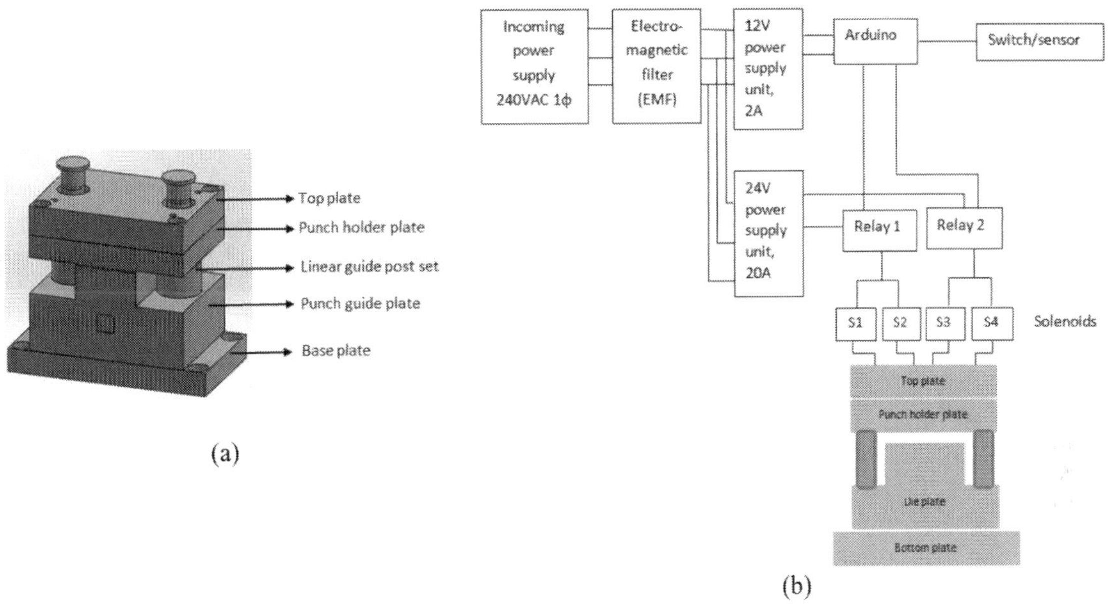

Figure2. (a) Machine and tooling structure and (b) Architecture of machine control system.

4. Finite Element Analysis

FE analysis was conducted to study each individual part and with the aim to focus on the deflection of the machine frame and moving structures(top and punch guide plate). The forces exerted simulates the forming force of the solenoid. A comparison studies were made to observe and analyzed deflection occurred on two different materials; Aluminium alloy 6061 and mild steel. The deflection tolerance used was 2.5μmlike the forming tolerance. By taking strip thickness as 50μm for example, the deflection preferably not be greater than 2.5μm.

4.1 Machine Frame Displacement Responses on Different Materials

Figure 3 and 4show the results of deflection of various machine frames on both aluminium alloys 6061 and mild steel.Four points of 200N forcewere exertedat the machine frame's top surface to simulate the actuationforce from the solenoids. The C-frame showed 0.1362mm and 0.04721mm of displacement for aluminium alloy 6061 and mild steel respectively. The deflection of this frame design was found larger from the allowable deflection tolerance. Meanwhile, the O-framestructure demonstrated1.13μm and 0.3835μm of deflection for aluminium alloy 6061 andmild steel respectively. Meanwhile, 4-column frame shows 12.46μm and 4.19μm deflection for aluminium alloy 6061 and mild steel while 2-column frame deflection was found to be at17.09m and 5.73mm for aluminium alloy 6061 and mild steel respectively.Table 1 summarizes the deflection results for all cases and materials. It can be concluded that the O-frame design produced stiffer structure characteristic for both materials with the least deflection which is still within the desired tolerance. In addition, both 2- and 4-column frames in mild steel materials also demonstrated higher strength and passed the design tolerance specification.

4.2 Upper Die Displacement Responses

Upper die plays an important role during forming process as it is the only moving component to transmit force from the solenoids to the lower die. Three different thickness for top and punch guide plateswere used; 15, 20 and 25mm. Deflection characteristic of different thickness of top plate on aluminium alloy

6061 and mild steel is shown as in Figure 5. Based on the results, the top plate with 15 mm thickness showed deflection at 7.20μm and 2.41μm,20mm thick, 3.27mm and 1.09mm for both aluminium alloy 6061 and mild steel respectively. Deflection characteristics for both top and punch holder plate were found to be within the tolerance for both materials. For 25mm thickness of top plate, the results demonstrated1.87μm and 0.624μmdeflection for aluminium alloy 6061 and mild steel respectively. Figure 6 presentsdeflection characteristic of different thickness of punch holder plate for both materials. The applied force simulates direct contact between the top and punch holder plate during actuation. Based on the results, punch holder plate with 15mm thickness produces 9.09μm and 3.07μm,for 20mm thickness, 2.47μm and 2.01μm, for 25mm, 0.83μm and 0.59μmm for aluminium alloy 6061 and mild steel respectively.

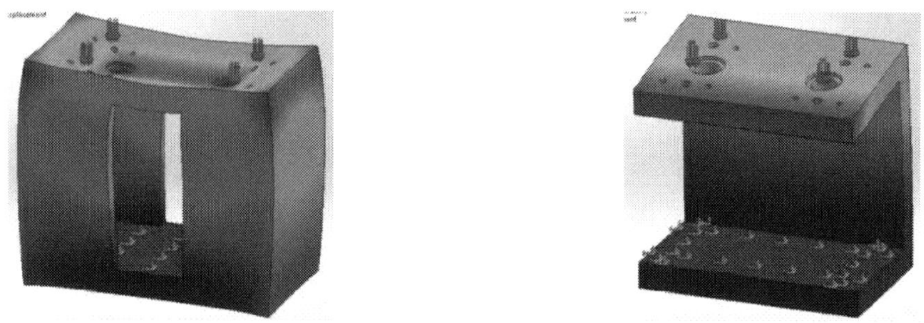

O-frame C-frame

Figure 3. Machine frame displacement responses on various machine design.

4-column frame 2-column frame

Figure 4. Comparison studies of machine frame deflection on aluminium alloy 6061 and mild steel.

Table 2summarizesthedeflection results for both top and punch holder plates. Both plate from mild steel material qualified the simulation evaluation which obey the design tolerance. However, due to good machinability characteristic of the aluminium alloy 6061 compared to the mild steel, this material has an advantage for the consideration. Based on the deflection results, 25mm and 20mm thick was selected for top and punch holder plate respectively.

4.3 Die Plate Displacement Responses

Figure 7 shows deflection occurred on the die plate for both materials. There is a part collection drawer/tunnel under the forming die which could weaken the structure. To determine the least deflection to accommodate the location of the part collectiondrawer, a study of the relationship between different positions of parts collection drawer's hole (denoted as distance d) and structure deflection was conducted.

Table 3summarizesdeflection on different drawer's position. The results indicated that the deflection occurred on both materials structure met the design tolerance. Therefore, changes in distance d did not contribute to the excess deflection of the machine's structure.

Table 1. Summary of machine frame deflection on aluminium alloy 6061 and mild steel.

Machine frame type	**Deflection (μm)**	
	Aluminium alloy 6061	Mild steel
C-type	136.20	47.20
O- type	1.13	0.38
4-column type	12.46	4.19
2-column type	17.09	5.73

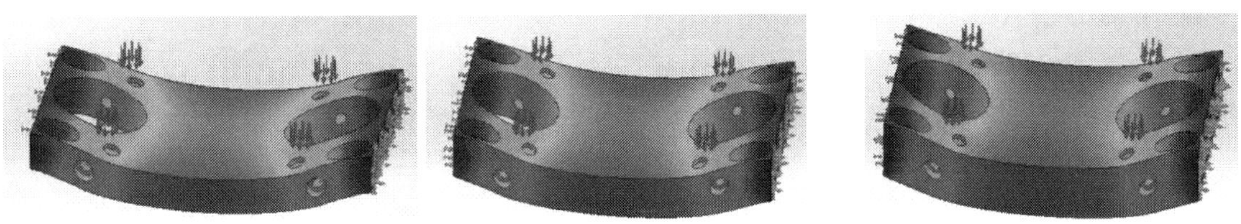

15mm 20mm 25mm

Figure 5. Comparison studies of top die deflection on both materials for different thickness

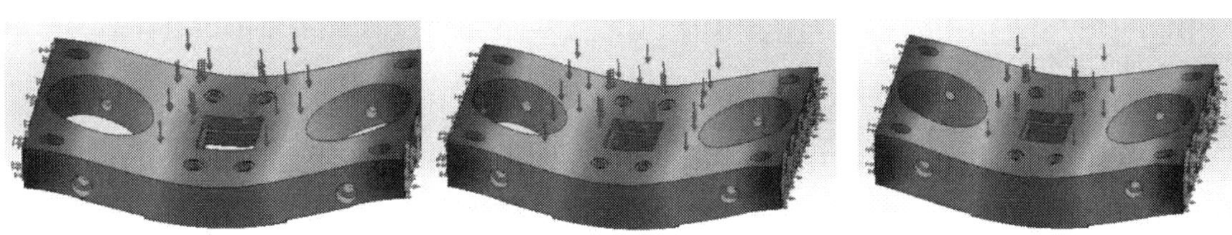

15mm 20mm 25mm

Figure 6. Comparison studies of punch holder plate deflection on both materials.

Table 2. Deflection ofthe upper die components

Upper die components	**Thickness (mm)**	**Deflection (μm)**	
		Aluminium alloy 6061	Mild steel
Top plate	15	7.20	2.41
	20	3.27	1.09
	25	18.69	6.24
Punch holder plate	15	9.09	3.07
	20	2.47	2.01
	25	8.33	0.59

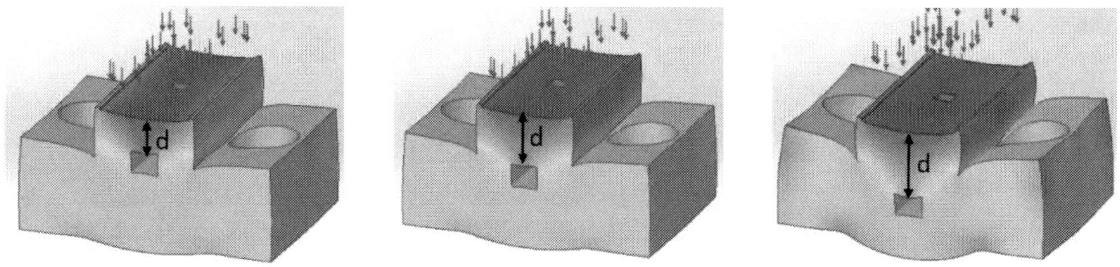

d = 20mm d = 30mm d = 40mm

Figure 7. Comparison studies of a die plate deflectionat different d distance.

Table 3. Summary of deflection for different drawer distance d from the top surface.

Drawer distance, d from top surface	Deflection (μm)	
	Aluminium alloy 6061	Mild steel
20	3.241	0.109
30	0.311	0.104
40	0.245	0.101

5. Conclusion

A design concept of a low energy machine system for a micro-sheet-forming was made based on the lowest sturdy O-frame type.Solenoids were used as a mean of actuation. Maximum deflection around wasfound to be between 2-3μm by using aluminium alloy 6061 with 25mm and 20mm thick on top and punch holder plate structures respectively.

References

[1] Akhtar Razul Razali, Yi Qin, *'A Review on Micro-manufacturing, Micro-forming and their Key Issues'*, Procedia Engineering, Vol. 53, (2013) pp 665-672.

[2] Akhtar Razali, *'A review on high-precision and high-speed devices for thin sheet metals feeding applications'*, Engineering Technology, Engineering Education, and Engineering Management, CRC Press 2015.

[3] Zhao, Jie; Yip, Arthur Lan Kuan; Qin, Yi; Razali, Akhtar; FEI, Zhi-cong, *'Machine and tool development for forming of polymeric tubular micro-components'*. Transactions of Nonferrous Metals Society of China (Elsevier Publication) Vol. 22, No. 2, 12.2012, p. s214–s221.

[4] Qin, Y., Razali, A., Zhou, M., Zhao, J., Harrison, C., and Wan Nawang, W.A. (2012). Dynamic Characteristics of a Micro-Sheet-Forming Machine System. Key Engineering Materials, Vols. 504-506, pp. 599-604.

[5] Akhtar Razali, *'Study of the feeding mechanism and influential parameters in strip feeding in micro-stamping – Positional overshooting'*, Engineering Technology, Engineering Education, and Engineering Management, CRC Press 2015.

[6] T.A. Chang, A.R. Razali, N.A.I Zainuddin, Y.W. Leong, *'Size effects in thin sheet metal forming'*, Materials Science and Engineering, 2015, Volume 100.

Structural modification of a steam turbine blade

M Heidari[1,2] **and K Amini**[1]
[1]Department of Mechanical Engineering, Islamic Azad University, Ahram Branch, Ahram, Iran

E-mail: heidarimd2@gmail.com

Abstract. Blades are significant components of steam turbines which are failed due to stresses arising from centrifugal and bending forces. The turbine blade has a number of geometrical variables that need to be considered at the design stage. Hence, this paper investigated a three-dimensional model of steam turbine blade with different lengths and thicknesses using finite element method. A three-dimensional model of blade was developed using a computer-aided design software. All materials were assumed linear, homogenous, elastic and isotropic. A 5 N widespread force was applied to the blade. The results of this study showed that longer blades are experienced higher maximum Von Mises stress and strain than shorter ones. The blade with the length of 400 mm and thickness of 20 mm experienced the lowest maximum Von Mises stress at 51 kPa. Furthermore, blade with the length of 400 mm and 600 mm experienced the lowest and highest strain at 3.07×10^{-6} and 4.3×10^{-6} respectively .In addition, thicker blades were undergone less maximum Von Mises stress and strain than thinner ones. Understanding stress and strain pattern in turbine blades provides useful knowledge which can be useful to estimate the fatigue in turbine blades.

Keywords: Steam turbine blade, Stress distribution, Strain, Finite element analysis

1. Introduction

Steam turbines are used in various types of power generation including nuclear power generation, coal-fired power generation, gas turbine combined-cycle power generation, and other power generation systems. Together, they provide the world with more than 60% of its electric power. Along with the extended use of renewable energy, providing electric power generated by steam turbines is about to increase [1]. In the thermal power station, the steam turbine is the prime mover which generates enough torque to produce power from generator. Steam turbine obtains its power by the adiabatic expansion of steam flow through the blades. Turbine has several parts which participate in the conversion of kinetic energy to mechanical energy [2]. Steam turbine blades are the heart of turbine which experiences the most intense static and dynamic conditions throughout their life span. Therefore, analysis of blades is compulsory to avoid any failure [3].

The blade system consists the turbine stages set in a series of subsequent units including rings of wide-ranging devices and rotor blades. Blade design combines the disciplines of thermo/aerodynamics and structural mechanics. While, the first ones strive for a design with maximum efficiency, second one is a restricting factor. In an optimization process, the mechanics provide the restrictions, while

[2] Corresponding author: M Heidari, heidarimd2@gmail.com

thermal/aerodynamics defines the objective function. Designing a blade is an iterative process with multiple levels of iterations. At first level, thermodynamical boundary conditions are calculated on the basis of black box models representing pipes, stage groups, turbine sections, valves, and etc. The second iteration level remarks a stage group with its geometrical data and considers aerodynamic variables, as e.g. structure mechanical variables e.g. natural frequencies and stresses. On further levels, individual optimizations, for blade fixations, are performed. Individual consists either just mechanical or just aero-dynamical variables [4, 5].

Blades are exposed to failure due to stresses arising primarily from centrifugal loads and bending forces related to the steam mass flow [6]. Furthermore, corrosion, causing crack initiation and propagation, is an significant failure mechanism in blades. This causes blades to be replaced or repaired and even probable re-design of elements [7]. In order to design a highly efficient steam turbine, it is essential to consider many design objective functions about the fluid dynamic performance (the smooth guidance of working steam, etc.) at the same time. Moreover, the steam turbine has a number of geometrical and topological variables (blade shape, the number of blades, number of stages, etc.) that also need to be considered at the design stage simultaneously [8].

Various blades exist which can be categorized based on their usage in the three turbine modules as high-pressure (HP), intermediate-pressure (IP) and low-pressure blades (LP). The first two turbine modules, HP and IP, are characterized by high temperatures and they contain small blades that should sustain small centrifugal forces, however, large blades contains bending forces due to impulse changes and high static pressure. They are equipped with T-roots assembled in tangential grooves around the rotor. The blades are bound to each other through integral roots and shrouds that ensure high stiffness of the blade row and as well as frictional damping to the structure. For the last stage blades, centrifugal forces are so high and bending by steam forces is approximately negligible. The largest rotating blades are equipped with fir-tree roots assembled in axial grooves. These roots include three or more lug pairs transmitting the forces portion-wise into the corresponding rotor lugs [5].

Richter et al. provided an overview of the structural design of modern steam turbine blades at Siemens power generation using the finite element code ADINAe [1]. The different types of blades including types of root shapes, root plus rotor steeple, different types of shrouds, and different types of snubbers were described in detail about their geometry and loading. For various analyses, a fatigue post-processor were implemented. Gajbhiye et al. performed vibration analysis using finite element analysis (FEA) technique on turbine blade and maximum allowable frequency was calculated. They showed that if the value of blade frequency rises beyond the natural frequency of blade the permanent damage accuses on blade [9]. Saxena et al. investigated failure in a more than 30 years old LMW design thermal power plant using FEA. The presented turbine blade failure was of LP turbine which has 8 stages. The results of each investigation were interpreted that leads finding the location of primary failure of the turbine blade, sequence and the root cause of its failure [10]. Segawa et al. developed a new rotor blade for steam turbine plants [11]. The new blade design optimized blade aerodynamics near the root section of the blade, hence reducing both profile and end wall losses. The new rotor blade was found to increase stage efficiency by about 0.3%.

This paper investigates a three-dimensional (3D) model of steam turbine blade with different lengths and thicknesses. A 3D model of steam turbine blade was developed using computer-aided design (CAD) software. Material properties and boundary conditions were applied. The models were analysed using finite element method (FEM) in order to obtain the Von Mises stress and strain in blades.

2. Materials and methods

A steam turbine blade was modelled using SolidWorks (Dassault Systemes SolidWorks Corp, USA) software. SolidWorks is a solid modelling CAD and computer-aided engineering (CAE) computer program. CAD model was prepared by taking dimensions of steam turbine blade from previous studies (Figure 1b). In order to model, the first stage steam turbine blade was chosen. The modelling of blade profile is an important task due to its direct effect on the simulation results. Cosmosworks (Dassault

Systems SolidWorks Corp, USA) was used for FEA using triangular mesh with the size of 2 mm. All materials were assumed linear, homogenous, elastic and isotropic. Elastic modulus and yield stress for turbine blade (chrome steel) were considered E=210 GPa and 585 Mpa respectively. Poisson's ratio (ν) and density were also set 0.3 and 7900 kg/m3 respectively [12]. Wide range amount of axial forces has been used for analysis of turbine blade, therefore in the present study a 5 N widespread axial force was applied to the blade which is relatively close to amount of previous studies [13]. The root of blade (base) was fixed in three directions (X,Y, Z) (Figure 2d). After applying material properties and meshing, next step was the solution using FEA. We selected the length (L) and thickness (t) of the blade as variable parameters for analysis. Therefore, 9 different cases was made as shown in table 1. The described models were used to obtain the Von Mises stress and strain in blades.

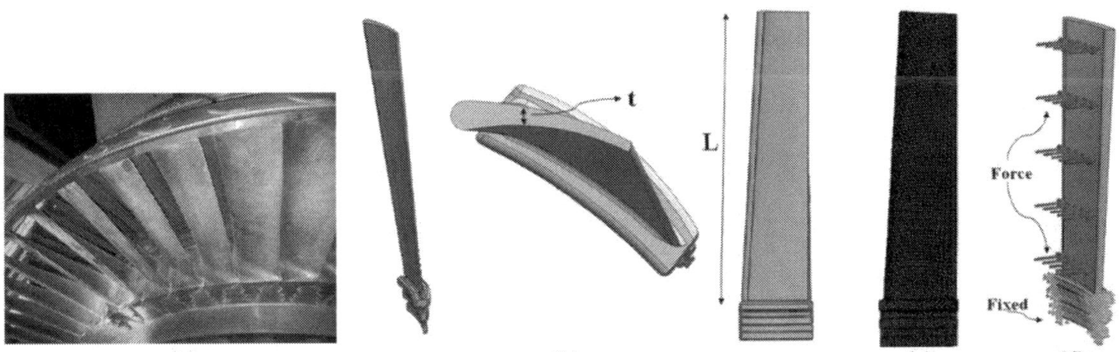

(a) (b) (c) (d)

Figure 1. Steam turbine blade (a) Real steam turbine blade (b) modelled blade (c) Meshed blade (d) Applied force and boundary conditions

Table 1. Variable parameters for turbine blade analysis

Case	Length (L)	Thickness (t)
1	400	10
2	400	15
3	400	20
4	500	10
5	500	15
6	500	20
7	600	10
8	600	15
9	600	20

3. Results

Results were obtained for nine various turbine blades. The different geometries affected the Von Mises stress and strain of blade (Figure 2). The highest stress was observed at the base of all blades with thickness of 10 mm. However, the lowest stress was seen in the tip of blades. Similarly, the most strain was seen in the base of blades, however, the tip of blades did not undergo any strain. Less stress was seen in the base of the blade with the length of 400 mm compared to blades with the length of 500 mm and 600 mm. As can be seen in Figure 3, blade with the length of 400 mm and thickness of 20 mm experienced the lowest maximum Von Mises stress at 51 kPa in comparison with all other models. However, blade with the length of 600 mm and thickness of 10 mm showed the highest maximum Von Mises stress at 237.73 kPa. For blade with the length of 400 mm, it was seen less maximum Von Mises stress in the blade with the thickness of 20 mm compared to the blade with the thickness of 15 mm (91.66 kPa) and 10 mm (161.34 kPa). Similarly, for the length of 500 mm, blade with the thickness of 10 mm experienced higher maximum Von Mises stress at 199.36 kpa than the blade with the thickness of 15 mm and 20 mm at 114.17 kPa and 63.43 kPa respectively. This trend was also observed for the blade with the length of 600 mm. Thicker blade (20 mm) showed less maximum Von Mises stress at 75.77 kPa than the blade with the thickness of 15 mm and 10 mm.

Blade with the length of 400 mm experienced the lowest strain at 3.07×10^{-6} while, the blade with the length of 600 mm showed the highest strain at 4.3×10^{-6}. For the blade with the length of 500 mm, strain decreased significantly from 3.8×10^{-6} for the thickness of 10 mm to $1.48 \times 10^{-6}$for the thickness of 20 mm. This trend was seen in a similar way for the blades with the length of 400 mm and 600 mm.

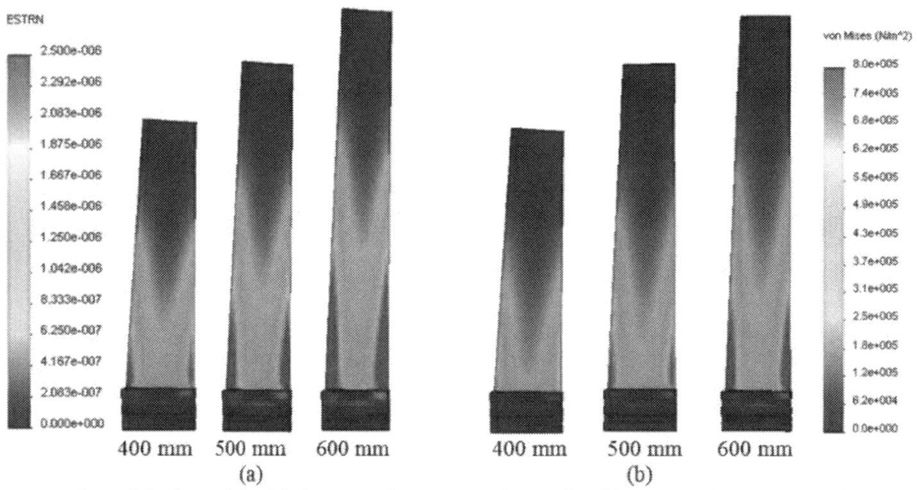

Figure 2. Various blades with thickness of 10 mm (a) Strain (b) Von Mises stress distribution

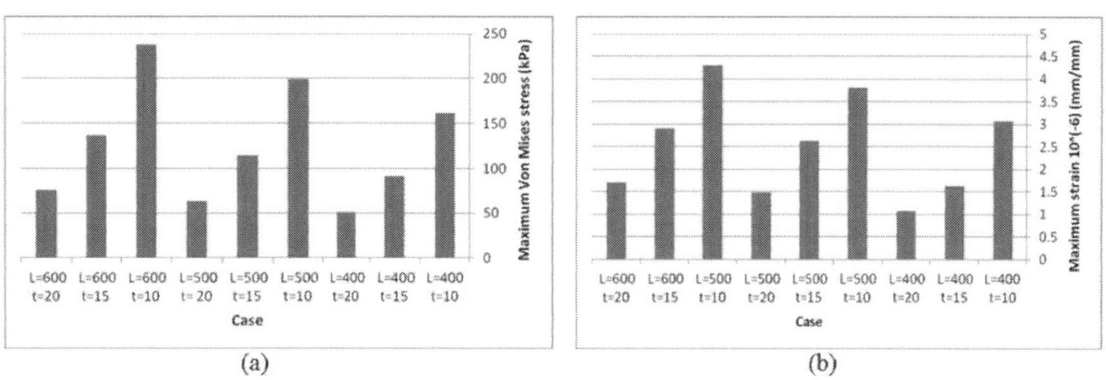

Figure 3. (a) Maximum Von Mises stress (b) Maximum strain for blade with different lengths (L) and thicknesses (t)

4. Discussion

In the present study, a steam turbine blade with various lengths and thicknesses were analyzed using FEM to obtain strain and Von Mises stress. Proper blade design with conservative stress and strain level is important for blade design. The major cause of malfunction in steam turbines is the failure of the blade. The failure of the blade may result catastrophic consequences both economically and physically. Therefore, the sufficient design of the turbine blade plays a significant role in the suitable functioning of the turbine [13]. A good design of the turbine blade consists the determination of steady loads acting on the blade and stressing due to them, determination of unsteady forces due to stage flow interaction, determination of geometric characteristics from gas dynamic analysis, determination of natural frequencies and mode shapes, and determination of dynamic forces and life estimation based on the cumulative damage fatigue theories [14]. However, in this study, we only concentrated on the effect of changing the geometry of blade on stress and strain.

Based on results, stress and strain were higher in base of blades than tip. The blade with the length of 400 mm and thickness of 20 mm experienced the lowest maximum Von Mises stress at 51 kPa in comparison with all other models. However, blade with the length of 600 mm and thickness of 10 mm showed the highest maximum Von Mises stress at 237.73 kPa. It presents preference of shorter and thicker blades to reduce Von Mises stress and strain compared with longer and thinner ones. Shorter blades have less surface and therefore, they are exposed to less force and deformation in comparsion with longer ones. Moreover, higher thickness causes blades to bend or deform less than thineer ones. In addition to this, blades with the length of 400 mm and 600 mm experienced the lowest and highest strain at 3.07×10^{-6} and 4.3×10^{-6} respectively. This is due to the fact that possibility of elongation in short blade are less than longer one. Understanding the stress and strain pattern in turbine blades provides useful knowledge which can be useful to estimate fatigue in turbine blades.

Studies on the state of stress and strain due to changing blade geometry are limited. Therefore, there is a call for further studies and investigations. Therefore, the present study introduced the modification of a steam turbine blade to decrease the factors that cause failure in blade including stress and strain using FEA. Length and thickness of turbine blade are design parameters which affect the longevity of the implant. The steam turbine blade is designed to work on high stresses so that it can withstand to high pressure and temperature. As it rotates at very high speed (approximate 25000 rpm) there are several chances of failure even though its advanced material composition and design [9]. As well as, the major cause of fatigue can be considered high stress and strain. Hence, designers need to observe significant consideration in order to reduce stress and strain in blades. Various limitations were witnessed in the present study. Material properties were assumed linear, however, considering nonlinear materials may be accommodated more reliable results. Centrifugal force and steam pressure are major forces which act on the blade. However, in this study steam pressure did not apply for analysis, which may or may not play a significant role in obtaining stress and strain changes in various blades.

5. Conclusion

The objective of this study was to modify a steam turbine blade using FEM. It was found that changing the geometry of turbine blade including length and thickness can be useful in decreasing Von Mises stress and strain. Results showed higher stress and strain at the base of blades rather than tips. It can be concluded that using shorter and thicker blades are better to reduce stress and strain, which may provide more durability.

References

[1] Senoo S, Asai K, Kurosawa A and Lee G 2013 Titanium 50-inch and 60-inch Last-stage Blades for Steam Turbines *Hitachi Review* **62(1)** 23-30

[2] Nurbanasari M A, Crack of a first stage blade in a steam turbine 2014 *Case Stud Eng Fail* **2** 54-60

[3] Shukla A and Harsha S P 2015 An experimental and FEM modal analysis of cracked and normal Steam Turbine Blade *Materials Today: Proceedings* **2** 2056-2063

[4] Jolanta B 2016 Redesign of steam turbine rotor blades and rotor packages- Environmental analysis within systematic eco-design approach *Energy Conversion and Management* **116** 18-31

[5] Christoph H R 2003 Structural design of modern steam turbine blades using ADINA *Computers and Structures* **81** 919-927

[6] Singh M P and Lucas G M 2011 Blade design and analysis for steam turbines The McGraw-Hill Companies Inc, New York united states of america.
[7] Arteaga C C, Rodríguez J A, Clemente C M, Segura J A and Urquiza G 2013 Estimation of useful life turbines blades with cracks in corrosive environment *Eng Fail Anal* **35** 576-589
[8] Shimoyama K, Yoshimzu S, Jeong S, Obayashi S and Yokono Y 2011 Multi-objective design optimization for a steam turbine stator blade using LES and GA. *J Comput Sci Technol* **5** 134-147.
[9] Bhupendra E G, Sachin V B and Kapil B S 2014 Vibration analysis of gas turbine blade profile blade using FEM technique and tool *I J Research in Advent Technology* **2(1)** 182-189
[10] Saxena S, Pandey J P, Solanki R S, Gaurav K G and Modi O P 2015 Coupled mechanical, metallurgical and FEM based failure investigation of steam turbine blade *Engineering Failure Analysis* **52** 35-44
[11] Segawa K, Shikano Y, Tsubouchi K and Shibashita N 2002 Development of a Highly Loaded Rotor Blade for Steam Turbines *JSME International Journal Series B* **45(4)** 881-890
[12] Tulsidas D, Shantharaja M and Bharath V G 2014 Life Estimation of a Steam Turbine Blade Using Low Cycle Fatigue Analysis *Procedia Materials Science* **5** 2392-2401
[13] Jabbari A A, Rai A K, Reedy P R and Dakhil M H 2014 Design and analysis of gas turbine rotor blade using finite element method, International *J Mechanical and Production Engineering Research and Development* **4** 73-94
[14] Soares C 2008 Gas turbine a hand book of air ,land and sea applications **2** Elsevier Inc, united states of america.

Responses of Multiple Nonlinear Tuned Vibration Absorbers under Harmonic Excitation

Abdullah S. Alsuwaiyan

Mechaniccal Engineering Depatment-Unayzah College of Engineering- Qassim University-Saudi Arabia. (Currently with Onaizah Private Colleges – Unayzah 51911 – Saudi Arabia).

asaoiean@qu.edu.sa

Abstract. In this paper, a system consisting of multiple perfectly tuned identical translational vibration absorbers, having both hardening and softening springs, attached to a main mass under harmonic excitation is considered. The existence of absorbers' synchronous and non-synchronous responses is checked. The method of averaging is employed to reach to the averaged autonomous equations of motion that describe the dynamics of the absorbers. A graphical method is then employed to check the existence of different responses of the absorbers. It is found that for absorbers with hardening springs, only one synchronous response of the absorbers occurs and no other responses take place. However, for the case of absorbers with softening springs, other responses were found to exist. These include multi-valued synchronous responses and a jump instability. These findings are in agreement with those of another study by the author where a similar system was considered using different approach.

1. Introduction

Tuned vibration absorber (TVA) is widely used to reduce excessive steady state harmonic vibrations of a machine or system that arise when the frequency of the excitation force that causes vibrations nearly coincides with the natural frequency of the machine or the system, i.e., at resonance situation. The invention of TVA goes back to 1909 by Herman Frahamm [1] (US Patent #989958, issued in 1911) and its use is considered to be one the most effective ways to achieve the goal of eliminating or reducing unwanted vibrations of engineering systems and their damaging effects. TVA has certain advantages over other vibration suppression methods. One of its big advantages is being external to the vibrating system, so no re-installation of equipment is necessary. Another big advantage of TVA is the fact that it does not require any external energy source to do its task. Vibration absorber, in its simplest form, consists of a mass attached to the vibrating system by an elastic element. The absorber mass and elastic element are chosen so that the natural frequency of the absorber is the same as the excitation frequency. In this situation, the absorber is said to be tuned to the excitation frequency. A damping element is usually added to the absorber in order to increase its effective bandwidth and to avoid system resonance and its damaging effects. Den Hartog [2] was the first to study TVA in depth and develop its theory which is very well established and can be found in any vibrations textbook, see for example [3].

Researchers considered nonlinear vibration absorbers in order to improve the absorbers' effectiveness. Samples of the research in this direction can be found in [4-10] and the references therein. Also, and due to some design constraints, designers may implement multiple linear or nonlinear absorbers instead of a single one to improve the absorbers task of reducing vibrations and meet certain design requirements. The reader is referred to references [11-15] as examples of the research in that direction.

In a previous work, Alsuwaiyan [16] considered a system of multiple perfectly tuned nonlinear vibration absorbers attached to a primary system under harmonic excitation. The stability of the absorbers' synchronous response, which is the response where all absorbers move together during operation, and the existence of other responses were checked. The absorbers considered are translational type with both hardening and softening identical springs having a similar cubic nonlinearity in each of them. A mathematical approach was implemented. It was shown that when the absorbers' springs are of hardening type, the absorbers' synchronous response is stable and no other responses exist under practical excitation force levels. However, when the absorbers' springs are of softening type, multi-valued responses were found to exist and jump instability was recorded.

In the current work, the same system considered in [16] is reconsidered. The stability of the absorbers' synchronous response and the existence of other responses are checked for both hardening and softening springs type. Here, a mathematical/graphical approach is implemented instead of the pure mathematical approach used in [16]. The findings of this work are to support their corresponding findings presented in [16].

2. Mathematical Model

A system of N translational vibration absorbers mounted on a primary mass, M, under harmonic excitation, as shown schematically in Fig. 1, is considered. The spring of each absorber has a linear spring stiffness, k_i and a cubic spring stiffness coefficient, α_i, *i.e.*, the restoring force of each absorber spring is: $F_{si} = k_i(x_i - y) + \alpha_i(x_i - y)^3$, where $i = 1, \ldots, N$.

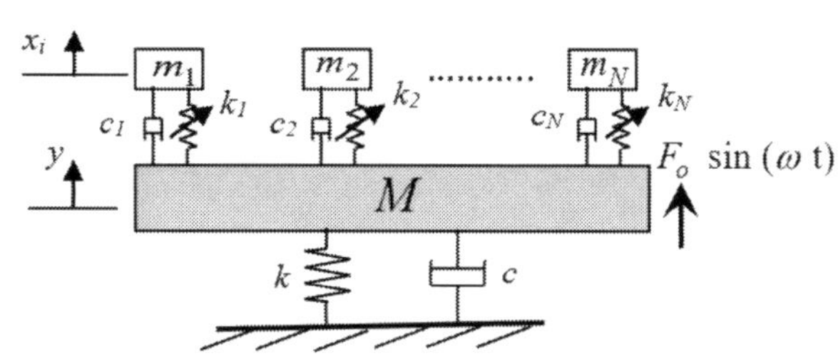

Fig. 1. Schematic diagram of N absorbers mounted on a primary mass, M

The equations of motion of this system are:

$$M\ddot{y} = F_o \sin(\omega t) - c\,\dot{y} - k\,y + \sum_{i=1}^{N}(k_a z_i + \alpha_i z_i^3 + c_a\,\dot{z}_i) \quad (1)$$

$$m_i\,\ddot{z}_i + c_a\dot{z}_i + k_a z_i + \alpha_i\,z_i^3 = -m_i\,\ddot{y} \quad (2)$$

Where,

$(\dot{\ }) = \frac{d}{dt}(\ )$, $(\ddot{\ }) = \frac{d^2}{dt^2}(\ )$, $z_i = x_i - y$, $i = 1, \ldots, N$. The absorbers are assumed to be identical meaning that that their masses, damping and spring stiffness coefficients are the same, *i.e.*, $m_i = m, c_i = c_a, k_i = k_a, \alpha_i = \alpha \quad \forall\, i \in [1, N]$.

Following the formulation procedure used in [16], one can reach to the following equation that describes the dynamics of each absorber uncoupled from the main mass dynamics:

$$\ddot{z}_i + \omega^2 z_i = \epsilon \left[-\tilde{\mu}_a \dot{z}_i - \tilde{\alpha}_a z_i^3 - f_o sin(\omega\, t) - \omega_a^2 \sum_{j=1}^{N} z_j \right] + O(\epsilon^2) \tag{3}$$

where, $\epsilon = \frac{m}{M}$, $\mu_a = \frac{c_a}{m}$, $\tilde{\alpha}_a = \frac{\alpha_a}{\epsilon}, \tilde{\mu}_a = \frac{\mu_a}{\epsilon}, \omega_a^2 = \frac{k_a}{m} = \omega^2 = \frac{k}{M}, \alpha_a = \frac{\alpha}{m}, f_o = \frac{F_o}{m}$.

The right hand side of Equation (3) is $(2\pi/\omega)$ periodic function of time. To put (3) into a form suitable for applying the method of averaging, the following invertible Vander Pol transformation can be used [17]:

$$\begin{pmatrix} u_i \\ v_i \end{pmatrix} = A \begin{pmatrix} z_i \\ \dot{z}_i \end{pmatrix} \tag{4}$$

where $A = \begin{pmatrix} \cos\omega\, t & -\frac{1}{\omega}\sin\omega\, t \\ -\sin\omega\, t & -\frac{1}{\omega}\cos\omega\, t \end{pmatrix}$

to obtain from (3)

$$\dot{u}_i = \frac{\epsilon}{\omega} \left[\tilde{\mu}_a \dot{z}_i + \tilde{\alpha}_a z_i^3 + f_o sin(\omega\, t) + \omega^2 \sum_{j=1}^{N} z_j \right] \sin\omega\, t + O(\epsilon^2) \tag{5}$$

$$\dot{v}_i = \frac{\epsilon}{\omega} \left[\tilde{\mu}_a \dot{z}_i + \tilde{\alpha}_a z_i^3 + f_o sin(\omega\, t) + \omega^2 \sum_{j=1}^{N} z_j \right] \cos\omega\, t + O(\epsilon^2) \tag{6}$$

Defining a scaled time τ as $\tau = \omega\, t$, and averaging these nonautonomous equations over one period in the scaled time, *i.e.*, (2π), gives the following associated autonomous averaged system to leading order in ϵ :

$$u_i' = \frac{\epsilon}{2} \left[-\tilde{\mu}_a\, u_i - \frac{3}{4}\, \tilde{\alpha}_a v_i (u_i^2 + v_i^2) + f_o - \sum_{j=1}^{N} v_j \right] \tag{7}$$

$$v_i' = \frac{\epsilon}{2} \left[-\tilde{\mu}_a\, v_i + \frac{3}{4}\, \tilde{\alpha}_a u_i (u_i^2 + v_i^2) + \sum_{j=1}^{N} u_j \right] \tag{8}$$

where $(\ )' = \frac{d}{d\tau}(\ )$

These equations can be expressed in polar coordinates by the usual transformation to polar coordinates, *i. e.*,

$$z_i = r_i sin(\tau + \theta_i), \;\; z_i' = r_i \cos(\tau + \theta_i) \tag{9}$$

where, $r_i = \sqrt{u_i^2 + v_i^2}$, $\theta_i = \arctan\frac{v_i}{u_i}$

Equations (7) and (8) are used to check the absorbers' synchronous response and the existence of other absorber responses in the next section.

3. Synchronous and Non-Synchronous Absorbers' Responses

To examine the existence of absorbers' synchronous and non-synchronous responses, the *N* absorbers are divided into two groups. The first consists of *Q* absorbers and the other consists of the remaining *N-Q* absorbers. The absorbers in each group vibrate in unison. The absorbers' displacements in the first group are (u_k , v_k), while the displacements of those in the other group are (u_l , v_l), where $k=1,2...,Q$ and $l=Q+1, Q+2,...,N$.

For the first group, equations (7) and (8) will be:

$$u_k' = \frac{\epsilon}{2}\left[-\tilde{\mu}_a\, u_k - \frac{3}{4}\,\tilde{\alpha}_a v_k (u_k^2 + v_k^2) + f_o - Q v_k - (N-Q) v_l\right] \tag{10}$$

$$v_k' = \frac{\epsilon}{2}\left[-\tilde{\mu}_a\, v_k + \frac{3}{4}\,\tilde{\alpha}_a u_k (u_k^2 + v_k^2) + Q u_k + (N-Q) u_l\right] \tag{11}$$

For the other group, equations (7) and (8) will be:

$$u_l' = \frac{\epsilon}{2}\left[-\tilde{\mu}_a\, u_l - \frac{3}{4}\,\tilde{\alpha}_a v_l (u_l^2 + v_l^2) + f_o - Q v_k - (N-Q) v_l\right] \tag{12}$$

$$v_l' = \frac{\epsilon}{2}\left[-\tilde{\mu}_a\, v_l + \frac{3}{4}\,\tilde{\alpha}_a u_l (u_l^2 + v_l^2) + M u_k + (N-Q) u_l\right] \tag{13}$$

To find the absorbers' steady state responses in both groups, the right hand sides of these equations, each equated to zero, should be solved for the absorbers displacements. This is done by employing the method used by Alsuwaiyan and Shaw [18] where they studied the existence and stability of the synchronous and non-synchronous responses in multiple centrifugal pendulum vibration absorbers. This method is as follows. One displacements pair is eliminated from two equations resulting in a two lengthy equations in the other pair. To be sure that all possible solutions are captured, the zero contours of the two equations are plotted in the displacement pair plane. The contour intersections are the solutions that represent the absorbers steady state responses. Absorbers' displacements in each group can then be transformed to amplitude and phase angle coordinates for physical interpretation of the results if needed.

It should be noted here that solutions satisfying (u_k , v_k) = (u_l , v_l) are the synchronous responses. In this case, the absorbers' vibration amplitude is related to the excitation force by [16]:

$$f_o = \sqrt{(\tilde{\mu}_a\, \omega\, r)^2 + \left(\frac{3}{4}\tilde{\alpha}_a\, r^3 + N\omega^2 r\right)^2} \tag{14}$$

To illustrate this procedure a numerical example is presented in the following section.

4. A Numerical Example

To compare the results of the current work with those in [16], the same example considered there is reconsidered here. It is a system of four absorbers (*N=4*) mounted on a main mass of *100 kg* with a natural frequency of *25 rad/sec*. The harmonic excitation force is applied to the main mass at resonance condition frequency and the absorbers are tuned to the excitation frequency with a mass and a linear spring stiffness of each absorber of *6 kg* and *3.75 kN/m* respectively. The nonlinearity coefficient in the absorbers' spring are taken to be *1 kN/m³* for hardening springs type and *-1 kN/m³* for softening springs type.

4.1 Hardening Springs Type

The above described procedure is performed on absorbers with $\alpha_a = 1\ kN/m^3$and the contours are plotted for different applied force levels in Figs (2) to (5). It is clear that the only one synchronous response exists and no other responses are present. This is in agreement with what has been concluded in [16].

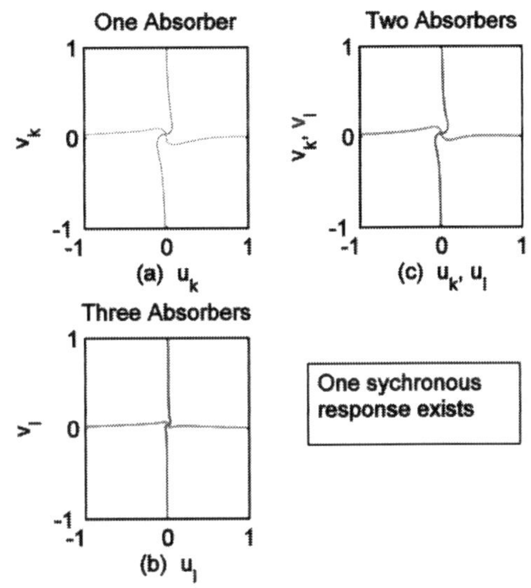

Fig. 2. $N=4$, $\alpha_a = 1\ kN/m^3$, $F_o= 0.5\ kN$; (a) and (b): $Q=1$, (c): $Q=2$

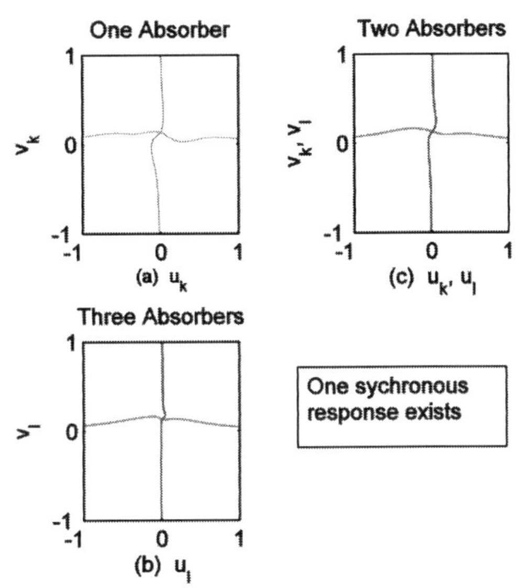

Fig. 3. $N=4$, $\alpha_a = 1\ kN/m^3$, $F_o= 2\ kN$; (a) and (b): $Q=1$, (c): $Q=2$

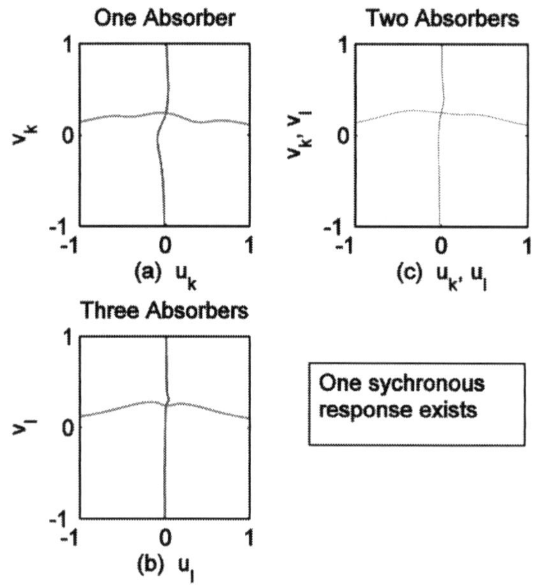

Fig. 4. *N=4* , $\alpha_a = 1\ kN/m^3$, *F_o= 4 kN*; (a) and (b): *Q=1*, (c): *Q=2*

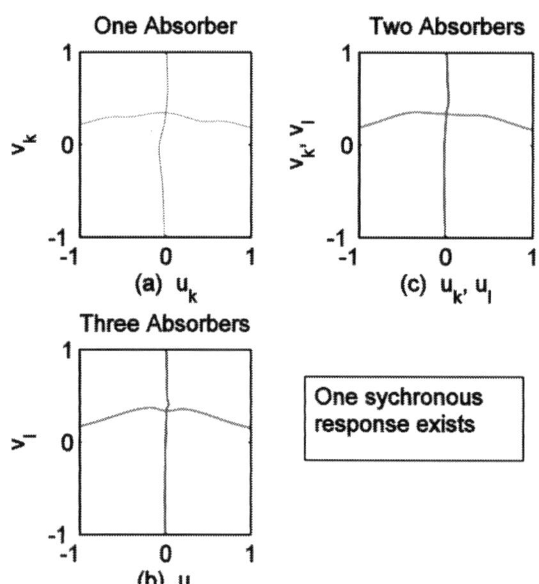

Fig. 5. *N=4* , $\alpha_a = 1\ kN/m^3$, *F_o= 6 kN*; (a) and (b): *Q=1*, (c): *Q=2*

4.2 Softening Springs Type

Contour plots for different applied force levels for absorbers with $\alpha_a = -1\ kN/m^3$ are shown in Figs (6) to (9). It can be clearly seen from Figs (6) to (8) that multi-valued synchronous responses (three here) exist and continue to exist with increasing applied force level. At a certain force level, two responses disappear and the third continues to exist as can be seen from Fig.(9). This means that jump instability takes place at the force level where the two responses disappear. This instability should be taken into consideration when implementing multiple absorbers with softening springs. Again, this is in agreement with the conclusions made in [16].

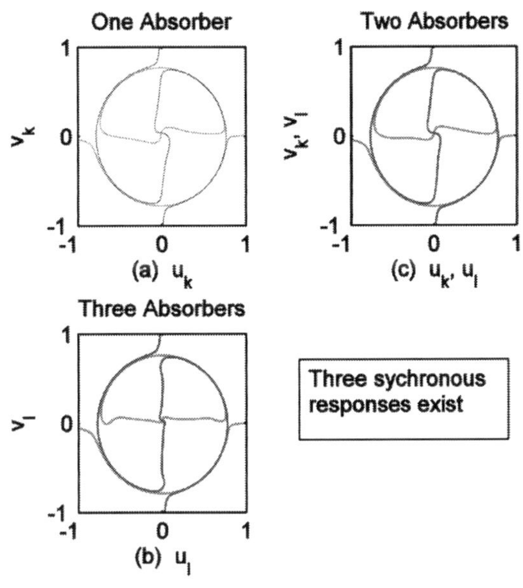

Fig. 6. N=4 , $\alpha_a = -1\ kN/m^3$, F_o= 0.5 kN; (a) and (b): Q=1, (c): Q=2

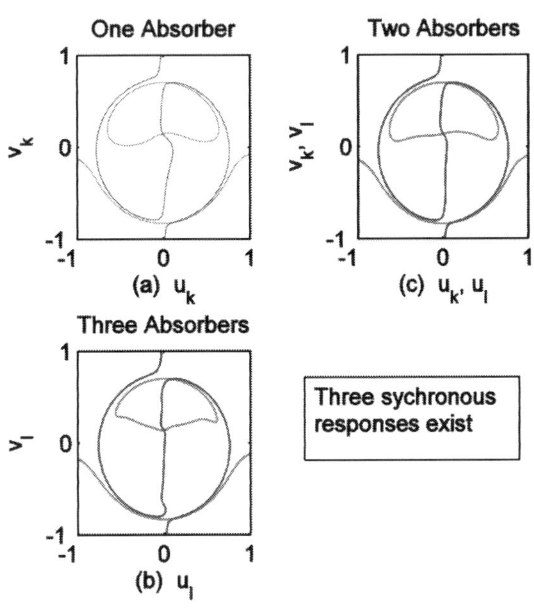

Fig. 7. N=4 , $\alpha_a = -1\ kN/m^3$, F_o= 2 kN; (a) and (b): Q=1, (c): Q=2

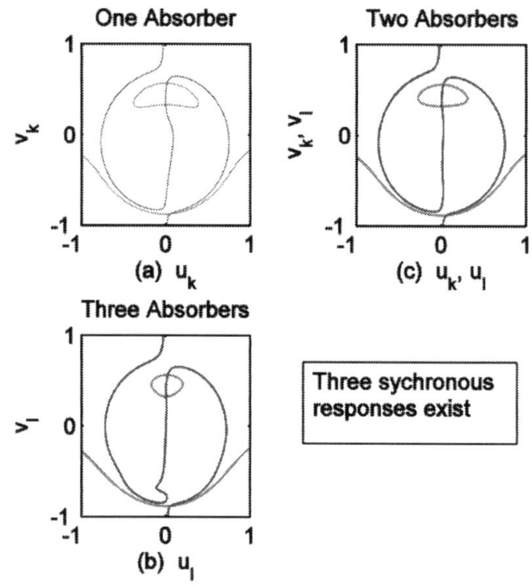

Fig. 8. $N=4$, $\alpha_a = -1\ kN/m^3$, $F_o= 4\ kN$; (a) and (b): $Q=1$, (c): $Q=2$

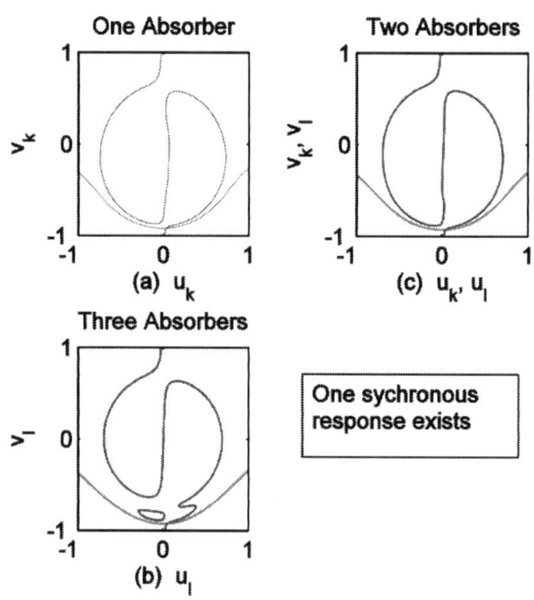

Fig. 9. $N=4$, $\alpha_a = -1\ kN/m^3$, $F_o= 6\ kN$; (a) and (b): $Q=1$, (c): $Q=2$

5. Conclusion

In this paper, the responses of perfectly tuned multiple identical translational vibration absorbers with both hardening and softening springs type were checked. For absorbers with hardening springs type, this work showed that only one synchronous response of the absorbers takes place. For absorbers with softening springs type, multi-valued synchronous responses and a jump instability were found to exist. These findings are in agreement with the recent work presented in [16] where the same dynamical system considered in this work was considered.

References

[1] H. Farham, "Device for Damping Vibration of Bodies," U.S. Patent #989958, 1911.
[2] J. P. Den Hartog, "Mechanical Vibrations", 4th edition, Dover Publications, Inc., New York, 1985.
[3] D. J. Inman, "Engineering Vibration", 4th edition, Prentice Hall, New Jersey, 2013.
[4] R. E. Roberson, "Synthesis of a nonlinear dynamic vibration absorber," J. Franklin Inst., vol. 254, pp. 205–220, 1952.
[5] J. Shaw, S. W. Shaw, and A. G. Haddow, "On the response of the nonlinear vibration absorber," Journal of Nonlinear Mechanics, vol. 24, pp. 281-293, 1989.
[6] J. B. Hunt, and J. C. Nissen, "The broadband dynamic vibration absorber," Journal of Sound and Vibration, vol. 83(4), pp. 573-578, 1982.
[7] A. S. Alsuwaiyan, "Design and Analysis of a Simple Nonlinear Vibration Absorber", IOSR Journal of Mechanical and Civil Engineering, vol.11(2), pp. 84-90, 2014.
[8] J. C.Nissen, K. Popp, and B. Schmalhorst, "Optimization of a non-linear dynamic vibration absorber," Journal of Sound and Vibration, vol. 99(1), pp. 149-154, 1985.
[9] I. N. Jordanov, and B. I. Cheshankov, "Optimal design of linear and non-linear dynamic vibration absorbers," Journal of Sound and Vibration, vol. 123(1), pp. 157-170, 1988.
[10] H. J. Rice, "Combinational instability of the nonlinear vibration absorber," Journal of Sound and Vibrations, vol. 108(4), pp. 526-532, 1986.
[11] L. A. Bergman, D. M. McFarland, J. K. Hall, E. A. Johnson, and A. Kareem, "Optimal distribution of tuned-mass dampers in wind sensitive structures," Proc.,5th Int.Conf.on Struct. Safety and Reliability, (ICOSSAR), ASCE, New York, N.Y., pp. 95-102, 1989.
[12] A. Kareem, and S. Kline, "Performance of Multiple Mass Dampers Under Random Loading," Journal of Structural Engineering, 121(2), pp. 348-361, 1995.
[13] S. S. Patil, S. B. Javehri, and C. G. Konapure, "Effectiveness of Multiple Tuned Mass Dampers," International Journal of Engineering and Innovative Technology, vol.1, pp.78-83, 2012.
[14] L. Zuo, and S. A. Nayfeh, "Optimization of the Individual Stiffness and Damping Parameters in Multiple-Tuned-Mass-Damper System," ASME Journal of Vibration and Acoustics, vol.127, pp. 77-83, 2005.
[15] T. H. Nguyen, I. Saidi, E. F. Gad, J. L. Wilson, and N. Haritos, "Performance of distributed multiple viscoelastic tuned dampers for floor vibration applications," Advances in Structural Engineering, vol.15(3), pp. 547-562, 2011.
[16] A. S. Alsuwaiyan, "Stability of the synchronous response of a system of multiple translational vibration absorbers", Arabian Journal for Science and Engineering, vol. 39, pp. 9071–9081, 2014.
[17] J. Guckenheimer, and P. Holmes, "Nonlinear Oscillations, Dynamical Systems, and Bifurcations of Vector Fields", 5th printing, Springer-Verlag, New York, 1997.
[18] A. S. Alsuwaiyan, and S. W. Shaw, "Non-synchronous and localized responses of systems of identical centrifugal pendulum vibration absorbers", Arabian Journal for Science and Engineering, vol. 39, pp.9205-9217, 2014.

IOP Conf. Series: Materials Science and Engineering **203** (2017) 012009 doi:10.1088/1757-899X/203/1/012009

Dehydration of Traditional Dried Instant Noodle (Mee Siput) Using Controlled Temperature & Humidity Dryer

K. A. Mamat[1], M.S. Yusof[1,4,]*, Wan Fauziah Wan Yusoff[2], M. Zulafif Rahim[1], S. Hassan[1], M Qusyairi. A. Rahman [3], M. A. Abd Karim[1]

[1]Faculty of Mechanical and Manufacturing Engineering, University Tun Hussein Onn Malaysia (UTHM) P.O. Box 101, Parit Raja, Batu Pahat 86400, Johor, Malaysia.

[2]Faculty of Technology Management and Business, University Tun Hussein Onn Malaysia (UTHM) , Parit Raja, Batu Pahat 86400, Johor, Malaysia.

[3] Center for Diploma Studies, University Tun Hussein Onn Malaysia (UTHM) , Parit Raja, Batu Pahat 86400, Johor, Malaysia.

[4] Tensei Innovative Solution, No. 51 Block C, Universiti Tun Hussein Onn Malaysia (UTHM) City Campus, Parit Raja, Batu Pahat 86400, Johor, Malaysia.

*mdsalleh@uthm.edu.my

Abstract. Drying process is an essential step to produce instant noodles. Yet, the industries especially Small and Medium Enterprises (SMEs), is seeking for an efficient method to dry the noodles. This paper discusses the performance of an invented drying system which employed heating and humidifying process. The drying system was tested using 30 kilogram of the raw noodle known as "Mee Siput". Temperature controlled system were used in the study to control the temperature of the drying process and prevent the dried noodles from damage by maintaining the temperature of lower than 80°C. The analysis shows that the system was drastically decreased the humidity from 80% to 40% just after 200 minutes of the drying process. The complete dehydration time of noodle has also decreased to only 4 hours from 16 hours when using traditional drying system without sacrificed the good quality of the dried noodle. In overall, the invented system believed to increase the production capacity of the noodle, reduce cost of production which would highly beneficial for Small Medium Industries (SMEs) in Malaysia.

1. Introduction

Malaysian local product of instant noodle or known as 'Mee Siput' is one of the local product that has a lot of potential to penetrate in a new business market. 'Mee Siput' is a traditional instant noodle produced by many rural entrepreneurs in southern region of Malaysia, particularly in Johor. The product comes in various types, shapes, formulations and contents, made from wheat, rice, buckwheat and starches [1]. 'Mee Siput' produced by dough mixing process combination of 6 basic ingredients; wheat flour, potato flour, Sodium Silicate, Sodium Bicarbonate, salt and water. The purpose of application of alkaline salt in noodle is to extend the lifespan by preventing the growth of fungus [1].

As the demand of the product not only came from Johor the products has been marketed throughout Malaysia including Singapore. Despite the increase in demand of the product from Singapore, the product has not yet produced in high quality and low production capacity. This is mainly due to the processing of the product that relied only on traditional method of drying.

The traditional drying processes and common practices for the traditional dried instant noodles are relied on air temperature ranges between 29°C and 35°C. As a result the noodle usually takes one to two days to complete the drying process in which it is depended on the weather condition. **Figure-1** illustrates the traditional drying method and dried noodle sample respectively.

Figure-1. 'Mee Siput' dried under the sun [2] and Sample of good quality Mee Siput dried by traditional method

Dried noodle can be produced using controlled drying process, in which the final moisture usually less than 14%. The noodle can be dried in a chamber where all parameters of temperature, humidity and ventilation are regulated [1]. This method is used to ensure long shelf life of the noodle. Drying is a classic process to preserve foods, which grant longer shelf life, making the food lighter and even smaller for storage and easy to be transported compared to wet product [3]. The process also believed to improve the food stability, reduce microbiological activity and minimizes physical and chemical changes during the food storage [4].

Elamin et.al in their study reported that the direct solar radiation as well as solar dryer are widely applied by small and medium scale industries for drying process [5]. During sun drying, inconsistency in temperature is a challenge in producing good quality of dried product. Nowadays, solar drying is the most popular method to preserve food based product. However contamination problem such as dust, insect, sand particles and soil is usually associated with this process. In fact, the drying time can be very long and inconsistent, as it is depended on the weather and temperature condition [1]. Controlling the temperature in between 60°C to 80°C are reported as able to produce high quality of dried noodles and pasta product [6][7][8]. N. Macmanus et.al stated that drying with regulated temperature of 55°C to 81°C and air velocity of 1.3 m/s to 2.51 m/s is able to remove 4.6 kg to 5.3 kg of water per day [9].

To solve the problem of inconsistent dried product quality and long duration of drying process, a new technology called a smart dryer machine was proposed. The machine was design purposely to reduce the drying time consumption and on the same time maintaining the hygienic drying condition. This paper presents the performance of the drying machine in term of its humidity reduction and temperature control capability

2. Methodology

2.1Materials

Raw material including wheat flour (Blue Key), tapioca starch (Cap ABC) and soda ash (American Natural Soda Ash Corporation, USA) are used to produce 30 kg noodle. All the materials were mixed with water using dough kneading machine. Small noodle making machine was used to cut the prepared dough sheet into noodle shape, similar to as used by T. Inazu [10]. The raw materials are physically examined to for quality inspection. To obtain uniform sample moisture distribution, the materials were kept at room temperature (25°C) for 15 minutes prior to experiment [11].

2.2Drying Experiment

The drying experiment was performed by the same procedure as used by Ridvan and Mustafa [12], using a custom made drying chamber as shown in **Figure-2.** The chamber (6 feet x 6 feet x 6 feet) was completely isolated with aluminium plate to maintain specific temperature and humidity during the drying process. Two liquid petroleum gas (LPG) heaters were used to supply constant heat through a connecting pipe to the drying chamber. The air blower was used to extract the heat from the heat pipe in which the constant blower air velocity of 2.62 m/s ensures uniform heat is distributed in the system. The chamber was equipped with a temperature controller (WILLHI, WH-7016) to trigger and turn off the blower when the temperature exceed the limit of 80°C. Dehumidifier (DeLonghi DN40G, Italy) was used to remove the moisture and vapour produced by the evaporation. The humidity value was monitored by a humidity sensor (MEXTECH TM-2 Digital Thermo-Hygrometer). An aluminium noodle tray (4X4 feet) from the dryer machine was used to fill the as-boiled noodle.

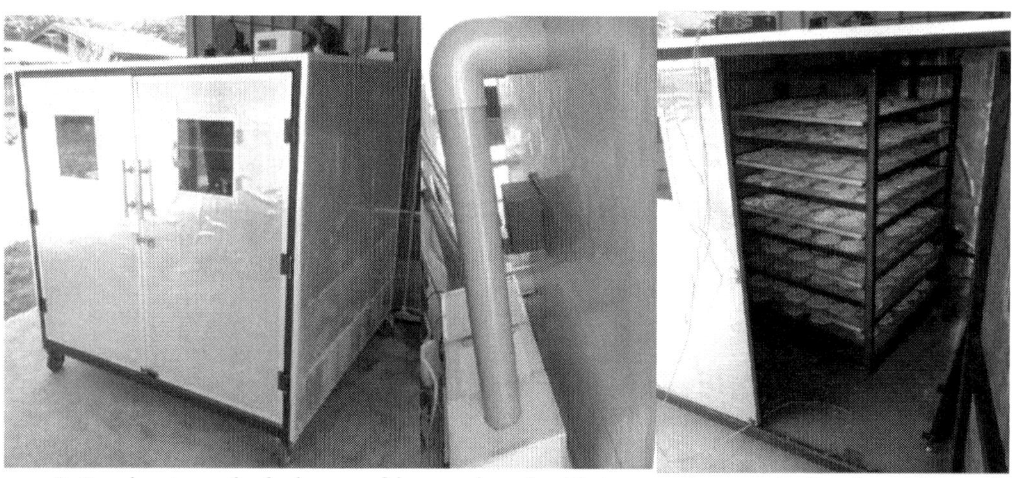

Figure-2. Semi-automatic drying machine equipped with temperature controller, humidity reducer mechanism

Total of 30kg samples (maximum chamber capacity) were loaded into the basket and rinsed for 15 minutes before separated into several trays and transferred into drying chamber. During the process, internal temperature of the chamber was controlled to maintain at temperature below 80°C. The samples were weighed after 5 hour to calculate the weight of water lost.

3. Result and Discussion

The first experiment was conducted using gas LPG heater only (without turn on the blower system) found that the rate of sample drying are different for each tray. It was found that the sample in the bottom tray (tray no. 7 to tray no. 10) need longer time to dry as they are continously facing lower temperature distribution in a range of 35°C to 40°C. The rest of tray that located at higher than tray no. 7 obtained almost equal temperature in the range of 65°C to 80°C. At longer processing period, poor heat circulation caused to the accumulation of heat at the upper portion of the chamber. It consequently produced high number of defective noodles which identified by dark brown in colour, high brittleness and shrank significantly due to the temperature higher than 80°C.

Drying temperature and heating rate will result in inconsistent moisture content as mentioned in reference [13]. Hence, the blower fan and dehumidifier blower system were installed as a heat circulation mechanism of the drying chamber. As a result, uniform drying rate was obtained at different noodle trays in the drying chamber. Contradict to the aformentioned phenomenon of initial experiment (without turn-on the blower), the samples from the bottom tray (no.10) had the same result in drying rate and weight with the middle tray (no.5) as well as the upper tray (no.1). This attributed by the uniform hot-dry air that flow vertically from the bottom exaust pipe of dehumidifier blower to the overhead suction pipe which provide continous and uniform hot- dry air at every level. In addition, the suitable velocity of hot air that blowed from the side wall was sucessfully delivered uniform hot air distribution in horizontal direction. In this case, to ensure uniform drying result for different tray in drying chamber, rearranging or swaping the trays it is not necessary. It was proved that the circulation system was successfully decreased the drying time from 7 hours to 5 hours (2 hour shorten) in drying process of 30 kg Mee Siput in a single drying machine.

As shown in **Figure-3,** the dryer was able to reduce the moisture content up to 50%. From the experiment, it was found that the humidity in drying chamber reduced with the increasing in processing time. Upon 200 minutes of drying operation, air humidity in drying chamber sucessfully reduced significantly from 80% to 42%. The humidity stagnated at that level for few times before dropped again when it is reaching 400 minutes of drying period. It can be expected that the humidity can be reduced further if the drying machine is built with more hermetic.

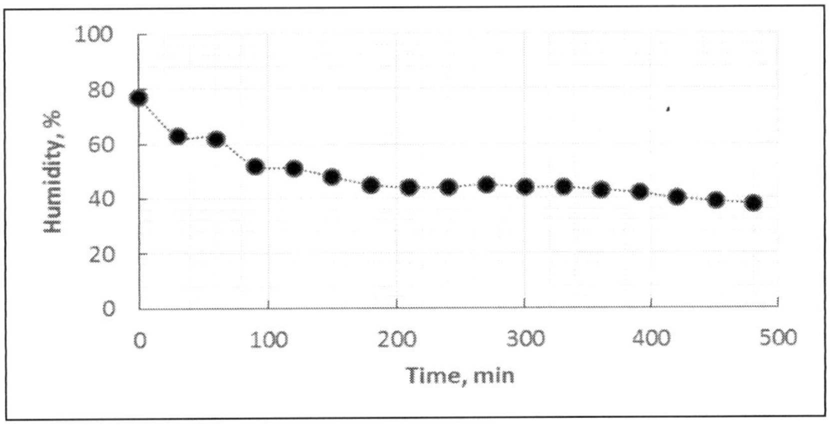

Figure-3. Relative humidity against drying time

As shown in **Figure-4**, a linear relationship of drying temperature with time was obtained. The temperature between 75°C to 80°C considered as the ideal temperature for drying without compromising the quality and texture of the noodle. The system that able to maintain the temperature of lower than 80°C was resulted to the production of good quality dried noodles as it will evade the texture damage that consequently increase the noodle brittleness and hardness. Higher the temperature of more than 80°C will make noodle turn into dark in colour. Investigation on the water removal ability found that average of 3.7 kg of water was removed after 5 hour of the drying process which is equal to 17.76 kg of water per day **(Figure 5).**

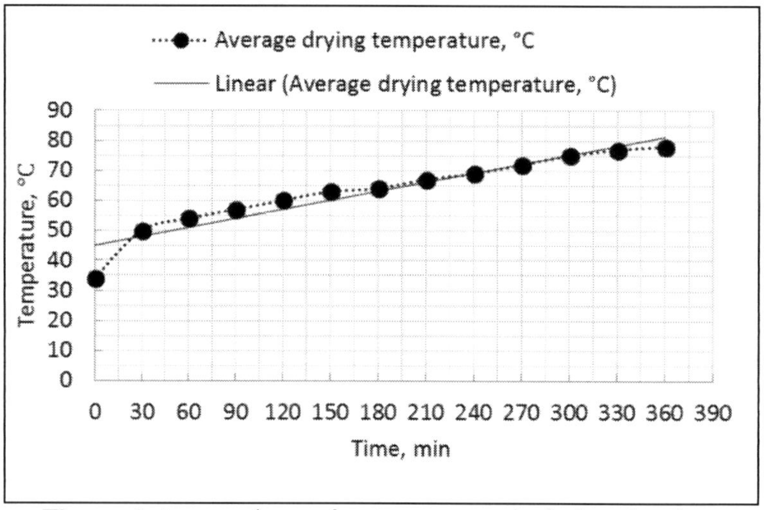

Figure-4. Average increasing temperature in drying chamber.

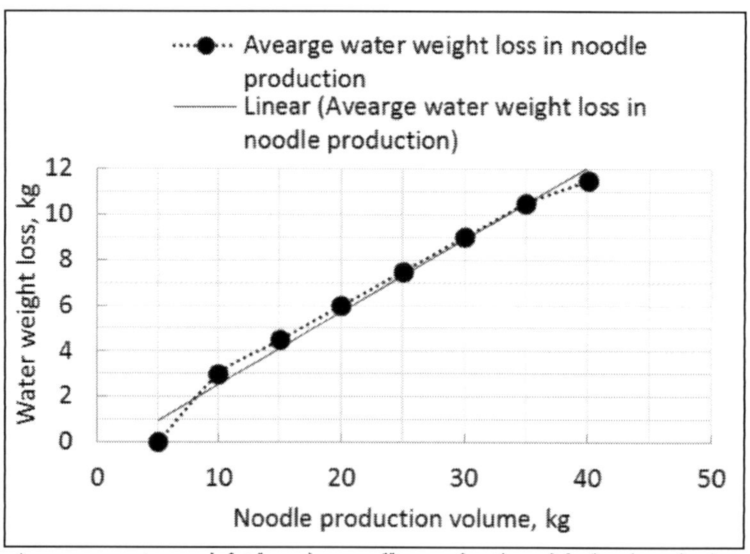

Figure-5. Average water weight loss in noodle production dried using drying machine.

Figure 6 shows the dried noodle after 5 hours of drying process. In overall, the invented system managed to perfectly dry the noodle. The samples produced shows no different with the original product produce by traditional method in term of its colour, odour and taste.

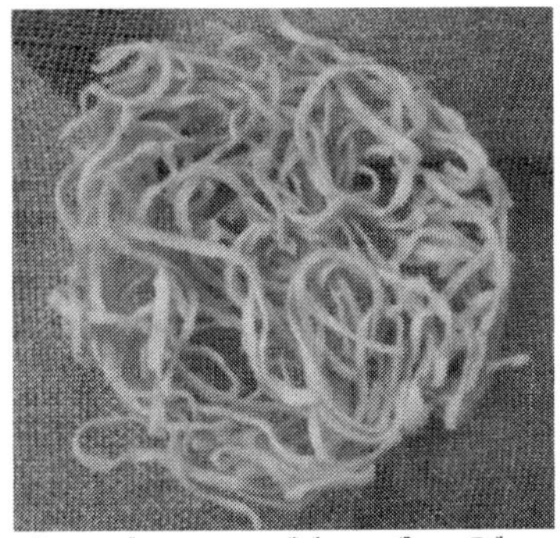

Figure-6. Result from dryer machine after 5 hour drying process.

4. Conclusion

Integrated system of temperature control and humidifier in the isolated chamber was successfully developed. The system was proved as able to provide high drying rate in which the efficient humidity reduction of 80% to 40% happened in a very short time. The system was also proved as able to evade the increasing in temperature of higher than 80°C. In comparison with conventional method, the drying time was successfully shortened from 16 hours to 5 hours. The significant improvement drying times was believed to provide an opportunity to the SMEs to increase their production capacity and on the same time maintain the require quality of the product. The system is also believed as applicable to be used for drying process of other product.

References

[1] B. X. Fu. (2008). Asian noodles: History, classification, raw material, and processing: Food Research International 41 888-902.
[2] F. Saat. "Fabrication and Production Studies of Mee Siput dryer machine for Small and Medium Industries (SME)," Degree Thesis, University Tun Hussein Onn Malaysia,2014.
[3] Ertekin C. & Yaldiz O. (2004). Drying of eggplant and selection of a suitable thin layer drying model from Journal of Food Engineering, 63, 349-359.
[4] L.Mayor, A.M.Sereno. (2004). Modelling shrinkage during convection drying of food material: a review from Journal of Food Engineering 61 373-386.
[5] Ertekin C. & Yaldiz O. (2004). Drying of eggplant and selection of a suitable thin layer drying model from Journal of Food Engineering, 63, 349-359.
[6] S.Villeneuve, P.Gelinas. (2006). Drying kinetic of whole durum wheat pasta according to temperature and relative humidity. Swiss Society of Food Science and Technology. doi:10.1016/j.lwt.2006.01.004.
[7] Dexter, J.E., Tkachuk, R., & Matsuo, R. R. (1984) Amino acid composition of spaghetti: Effect of drying condition on total and available lysine. Journal of Food Science 49, 225-228.
[8] De Stefanis, E., & Sgruletta, D. (1990). Effect of high temperature drying on technological properties of pasta. Journal of Cereal Science, 12, 97-104.
[9] N.Macmanus, and C.M. (2009). Effect of Drying Temperature and Drying Air Velocity on the Drying Rate and Drying Constant of Cocoa Bean from CIGR Ejournal. Manuscript1091. Vol. XI.
[10] T. Inazu, K. Iwasaki. (2000) Mathematical Evaluation of Effective Diffusivity in Fresh Japanese Noodle (Udon) by Regular Regime Theory. Journal of Food Science-Vol. 65, No.3
[11] S.Abasi, S.M.Mousavi, M.Mohebi and S.Kiani. (2009) Effect of Time and Temperature on Moisture Content, Shrinkage, and Rehydration of Dried Onion from Iranian Journal of Chemical Engineering Vol. 6, No. 3 (Summer), IACHE

[12] R. Kucuk, M. Ozilgen .(1997) Predicting Drying Behaviour of Spaghetti in a Continuous Industrial Dryer With The Model Determined In a Lab Scale Bath System. Journal of Food Processing and Preservation 21 (1997) 245-256.

[13] V. K. Jindal, T.J. Siebenmorgen. (1987). Effect of Oven Drying Temperature and Drying Time on Rough Rice Moisture Content Determination from article of the ASAE (Vol. 30, No. 4, pp. 1185-1192).

Petroleum sludge treatment and reuse for cement production as setting retarder

A K Aeslina[1*] and B Ali[1]

[1]Department of Water and Environmental Engineering, Faculty of Civil and Environmental Engineering, University Tun Hussein Onn Malaysia, 86400 Parit Raja, Batu Pahat Johor, Malaysia

[2]Division Recherche & Développement-Amont SONATRACH/Direction Centrale Recherche & Développement, Avenue 1st November 35000 Boumerdès, Algeria

*aliptbe2008@gmail.com

Abstract. Petroleum sludge is a dangerous waste that needs to be treated to avoid any contamination of soil and groundwater due to its disposal. As an attempt to treat this waste, it has been incorporated into cement production as substitution for gypsum. As results, 5% of petroleum sludge has shown effective results and could play the same role of gypsum in delaying the flash setting of cement clinker.
Keywords: Petroleum sludge, cement, setting time, gypsum replacement, heavy metals.

1. Introduction

The unavoidable outcome to the different petroleum production processes is the generation of a considerable amount of waste called petroleum sludge (PS). This last is referring to a thick, viscous [1] intractable mixture encountered during the crude petroleum exploration, production, transportation, storage and refining processes [1, 2]. The PS is constituted predominantly of water, solids of the crossed geological formations during the drilling operation and hydrocarbons that are deduced from both the drilled petroleum itself and the substances used for the drilling operation. According to Resources Conservation and Recovery Act (RCRA), sludge is a hazardous waste [3] that affects the environment and the living things if it is disposed improperly. However, abundance, Intractability and harmful impacts of PS have made it a source of danger that attracted various types of treatment methods like centrifugation, solidification/stabilization, solvent extraction, landfarming, pyrolysis, biodegradation, photocatalysis, ultrasonic treatment, incineration and others. Researchers have concluded that it is ineffective to implement a typical method to get rid fully of the PS danger without any side effects. However, all the methods are associated with advantages and disadvantages that may prompt implementing such method for some purposes and not for others. Incineration is a promised treatment method that has attracted many researchers since long time ago for the treatment of wastes generated from industries. Nevertheless, this method is much expensive (130- 150 USD/ ton) and needs further treatment for the ash produced [4].

As a new sustainable treatment, in 2006, MVW Lechtenberg & Partner have concluded that the PS is feasible to be implemented into cement plants, but as a fuel only [4]. The study has not shown a complete data about the whole process such as the raw materials substitution, destination of ash, impact of the project and others.

While gypsum is a main component in cement production since it has a role to regulate the setting time, the XRD of PS shows that is contains a considerable amount of anhydrite, which is the main component of gypsum. Thus, it may play the same role of gypsum in cement production.

The current study aims to incorporate PS into cement production as gypsum replacement by examining the impact of this waste on the setting time.

2. Materials and methods

2.1. Materials Collection

2.1.1. *Clinker.* The clinker was collected from SCAEK (Cement Company of Ain EL- Kebira) Subsidiary of the group GICA (Industrial Group of Algerian Cement) situated in Ain El-Kebira, Setif, Algeria. Then, it was 105°C oven dried until it lost its moisture, grinded to pass through 1mm pore diameter metallic sieve and stocked in a polyethylene bag prior to be analyzed.

2.1.2. *Gypsum.* Gypsum has been collected from the career of The GRANU EAST Company Subsidiary also of the group GICA based in the municipality of Beni Fouda, State of Setif, Algeria. This gypsum, which is also supplied to SCAEK, was dried under a temperature of 50°C, grinded to pass through 1mm pore diameter metallic sieve and stocked in a polyethylene bag prior to be analyzed.

2.1.3. Petroleum Sludge. Petroleum Sludge was collected from the oil drilling field of Hassi Messaoud, Algeria. It was then burned in an industrial kiln to eliminate the organic matter and to measure the gases emitted through burning. After that, it was also grinded to pass through 1mm pore diameter metallic sieve and stocked in a polyethylene bag prior to be analyzed.

All materials: clinker, gypsum and petroleum sludge were analyzed by using the X-Ray Fluorescence (XRF) in order to identify their chemical components.

2.2. Petroleum Sludge Analysis

2.2.1. *Loss of Ignition (LOI).* This test was conducted according to ASTM D7348 by weighing 2g of petroleum sludge in a crucible, placing it in a furnace with an air atmosphere and heating it under a temperature of 950°C for 2h. The fired sample would be then cooled down to room temperature in a desiccator and weighed again. The weight loss associated with firing the sample is known as the loss of ignition (LOI) [5].

2.2.2. *Leaching.* This test was conducted by extraction using Toxicity Characteristics Leaching Procedure (TCLP) according to EPA Method 1311 and analysis using inductively coupled plasma mass spectrometry (ICP-MS). For that, 5g of PS was extracted with 20:1 liquid to solid ratio in an acetate solution in 2L bottles and was rotary agitated at 30 rpm for 18 hours using the rotary system [6]. After that, the solid and liquid phases were separated by filtration through 0.45-mm-pore-size membrane filters. The pH was measured, then; the extract was acidified to 2% HNO_3 prior to analysis for metals using ICP-MS. Then, the heavy metals quantities in the solutions obtained would be compared with the regulatory cited by the United States Environmental Protection Agency (USEPA).

2.3. Cement Preparation

Different percentages of clinker, gypsum and PS have been determined according to cement production requirement obtained from the production unit of SCAEK in order emulate the cement produced there. As they use 2.5% to 5% of gypsum added to 97.5% to 95% of clinker, respectively,

1%, 3.5% and 5% of petroleum sludge have been chosen as replacement of 20% and 100% of gypsum, respectively. Table 1 represents the samples percentages. It should be noted that the choice of these percentages is based on the percentage of SO_3 in cement that should not exceed 2.5% [7].

Table 1. Samples preparation

Samples	Petroleum Sludge (%)	Gypsum (%)	Clinker (%)
Reference	0	5	95
1	1	4	95
2	3.5	0	96.5
3	5	0	95

Each sample has been prepared separately and blended within a grinder for 30 minutes. The blending operation should be continued until 95-97% of the obtained cement could pass through 90μm pore size diameter metallic sieve in order to obtain a finesse of 3700 -4200 cm^2/g. After that, the samples were kept in polyethylene bags to prevent any source of moisture prior to be tested.

2.4. Setting Time

This test has been performed through two steps: consistency and Vicat needle

Consistency was performed by mixing 500g of dried cement with a measured amount of distilled water, then; the paste was put quickly into the conical Vicat ring, completely filled by leveling it with the upper tight edge of the ring and the excess paste was removed by using a trowel [8]. After that, the ring was centered under the rod which was brought in contact with the surface of the paste, the set screw was tightened to the upper zero mark of the scale, and then; it was released immediately to read the value indicator below the origin surface of the paste. For normal consistency, a value of 6 ± 2 mm should be obtained after making trials by mixing fresh cement each time. From mixing of paste until releasing of the rod, 30 seconds should not be exceeded [8].

Time of Setting Determination was effecting by putting the previous prepared paste in a mould and remained in a moist cabinet in the automatic Vicat apparatus (adapted to a computer) prior to start measurements. After 30 minutes, the automatic Vicat was starting penetrating every 15 minutes until a penetration of 25 mm was obtained. The elapsed time between the initial contact of cement and water and the penetration of 25 mm is the initial time setting [9]. But the final setting time was the measure when the needle did not mark the specimen surface with a complete circular impression [9]. The calculation of initial and final setting time was made automatically by the apparatus and the results were shown on the screen of the computer.

3. Results and discussion

3.1. Chemical Analysis

The chemical components of clinker, gypsum and petroleum sludge are shown in table 2.

Table 2. Chemical compositions of clinker, gypsum and petroleum sludge

Materials \ Components	SiO_2	Al_2O_3	Fe_2O_3	CaO	MgO	SO_3	K_2O	Na_2O	PAF	TOT
Clinker	21.15	4.34	5.70	66.49	1.33	0.18	0.08	<0.05	0.32	99.59
Gypsum	10.74	3.35	1.30	27.90	3.83	29.51	0.44	0.93	21.76	99.76
Petroleum Sludge	5.47	1.29	0.56	25.05	3.47	38.41	0.27	3.29	20.80	100.00

It is seen that petroleum sludge and gypsum have high content of Sulfur trioxide (SO_3) which is in charge of delaying the cement setting by controlling the early reactions of tricalcium aluminate [10]. Indeed, without existence of a required amount of SO_3 in cement, it can lead to flash setting [11]. GICA produces cement with SO_3 requirement between 2 and 5%, and in this study the XRF showed that the clinker has only 0.18 %of SO_3, thus, it should be regulated with a sufficient and adequate amount of gypsum and/ or petroleum sludge.

3.2. The loss on ignition (LOI)

LOI of petroleum sludge was made by a muffle furnace at a temperature of 900°C; the results gave an average loss value of 20.80%. This value describes water, the organic content like onentalkanes, aeromatics, asphaltenes and resin [51] and even some percentages of carbonate matters. It may also contain volatile organic carbons (VOCs) and semi-volatile organic carbons (SVOCs) like polycyclic aromatic hydrocarbons (PAH) [12, 13]. Thus, this value should be eliminated prior to use the petroleum sludge as gypsum replacement.

The ash (or mineral matter) content is the value of the residue remaining after the loss on ignition is achieved, and therefore the result is expressed as ash%= 100% - LOI = 100-20.80= 79.20%. Thus, the ash content of the petroleum sludge is 79.20%. This amount of ash may be beneficial as replacement for gypsum in cement production.

3.3. Leaching

The analysis of the leachate by using the ICP-MS is presented in table 3.

Table 3. Heavy metals content of petroleum sludge

Elements	Mn	Ni	Zn	Fe	Cu	Pb	Cd	Cr
Concentration (ppm)	25.77	0.34	0.20	37.40	3.43	149.02	0.44	0.48

This analysis shows that the heavy metals have not exceeded the regulatory limits for cement for all the elements except for lead (Pb), which is 149.02ppm compared to the maximum percentage in cement, which is 75ppm Table 4 shows the maximum concentrations of some heavy metals from a number of cement plants throughout the world. Thus; incorporation of petroleum sludge in cement plants should take in consideration this value to prevent any danger comes from this waste. In the current study, the maximum amount of PS incorporated would not excess 50% of any raw material used for cement production.

Table 4. Maximum concentrations of some heavy metals from a number of cement plants throughout the world

Heavy metals	Maximum concentrations (ppm)
Manganese (Mn)	2366
Nickel (Ni)	129
Zinc (Zn)	321
Iron (Fe)	N/A
Copper (Cu)	N/A
Lead (Pb)	**75**
Cadmium (Cd)	1.12
Chromium (Cr)	422

3.4. Setting time

Table 5 shows the initial and the final setting times for all samples.

Table 5. Initial and final setting times for the different samples

Samples (CK/GY/SL)	Initial setting time	Final setting time
Reference	1h55min	4h05min
1	1h55min	4h00min
2	2h10min	3h45min
3	> 2h10min	> 3h45min

It is noticed that all samples respond to the minimum initial setting time for cement, which is 75 minutes [14].

Sample 1, where 20% of gypsum was substituted by PS, showed the same results of initial setting time compared to the reference, and a slight increase (5 minutes) in the final setting time. These results are explained with the non-negative impacts of PS in substituting gypsum at lower percentage.

For sample 2 and sample 3, substitution of gypsum with the different percentages showed an increase in initial setting time and a decrease in the final setting time compared to the reference sample, but the differences are not that much. These results can't be explained only with the efficiency of using petroleum sludge in slowing down the early hydration of tricalcium aluminate (C_3A) and preventing the flash setting of the cement clinker [10].

4. Conclusion

From the above results, cement plants could offer a sustainable disposal of petroleum sludge could be treated within by offering. The main conclusions are expressed as follow:

1. Petroleum sludge contains a high amount of Sulfur trioxide (SO_3) that helps in delaying the flash setting once added to the cement clinker, thus, it is worth to be used as a substitute for gypsum.
2. Petroleum sludge plays the same role of gypsum as setting retarder.
3. The heavy metals constitution of petroleum sludge even though they are lower than the acceptance limits except the lead (Pb), but the incorporation percentage should be controlled to prevent any source of danger comes from this heavy metal.

Acknowledgment
The authors want to thank C.E.T.I.M laboratories, Algeria and UTHM University for their support to execute this work.

References

[1] Prakash V, Saxena S, Sharma A, Singh and Singh S K 2015 *Treatment of Oil Sludge Contamination by Composting* J. Bioremed & Biodegrad **6**(3) 1

[2] Organization of the Petroleum Exporting Countries (OPEC) 2015 *Monthly Oil Market Report Feature article: Review of 2015, outlook for 2016* (Electronic Materials) Vienna, Austria From http://www.opec.org/opec_web/static_files_project/media/downloads/publications/MOMR_December_2015.pdf

[3] United States Environmental Protection Agency (USEPA) 2008 *Hazardous waste listings* (The United States) from http://www.epa.gov/osw/hazard/wastetypes/pdfs/listingref

[4] Lechtenberg D 2010 *Waste management and cement industry in Arab countries* Cement & Building Materials Review **39**

[5] American Society for Testing and Materials (ASTM) 2007 *ASTM D5468-02 Standard Test Method for Gross Calorific and Ash Value of Waste Materials* (West Conshohocken, The United States)

[6] USEPA 1992 *Method 1311–Toxicity Characteristic Leaching Procedure (TCLP)* (The United States)

[7] Algerian Institut of Standardization (Fr. IANOR) 2006 *NA 443 Cement for work in highly aggressive environments* (frensh language) From : http://fr.calameo.com/read/001496329655dc7795f49

[8] American Society for Testing and Materials (ASTM) 2004 *C 187-04 Standard Method of Test for Normal Consistency of Hydraulic Cement* (West Conshohocken, The United States)

[9] American Society for Testing and Materials (ASTM) 2004 *C 19 Standard Test Method for Time of Setting of Hydraulic Cement by Vicat Needle* (West Conshohocken, The United States).

[10] Amin A H 2004 *Influence of the SO_3 Content of Cement on the Durability and Strength of Concrete Exposed to Sodium Sulfate Environment* (Master thesis) (University of South Florida, The United States)

[11] Mindess S, Young J F and Darwin D 2003 *Concrete, 2nd ed.* (Upper Saddle River, NJ: Prentice Hall Pearson Education, Inc.)

[12] Mishra S, Jyot J, Kuhad R C and Lal B 2001 *In situ bioremediation potential of an oily sludge-degrading bacterial consortium.* Curr. Microbiol **43**(5) 328–5

[13] Bach Q Kim S Choi S & Oh Y 2005 *Enhancing the intrinsic bioremediation of PAH-contaminated anoxic estuarine sediments with biostimulating agents* J. Microbiol **43**(4) 319

[14] European Committee for Standardization 2000 *SN EN 197-1 Cement - Part 1: Composition,*

IOP Conf. Series: Materials Science and Engineering **203** (2017) 012010 doi:10.1088/1757-899X/203/1/**012010**

specifications and conformity criteria for common cement (Stassart Street: Brussels) from: http://www.holcim.az/fileadmin/templates/AZ/images/Technical_Solutions/EN-197-1_en.pdf

IOP Conf. Series: Materials Science and Engineering **203** (2017) 012011 doi:10.1088/1757-899X/203/1/012011

Leaching of Heavy Metals Using SPLP Method from Fired Clay Brick Incorporating with Sewage Sludge

Aeslina Abdul Kadir*[1], Nurul Salhana Abdul Salim[1], Noor Amira Sarani[1], Nur Aqma Izurin Rahmat[1],
[1]Faculty of Civil and Environmental Engineering, Universiti Tun Hussein Onn Malaysia, 86400 Parit Raja, Batu Pahat Johor, Malaysia

E-mail: *aeslina@uthm.edu.my, nurulsalhana@gmail.com, aqizin@gmail.com, nramira1987@gmail.com

Abstract. Sewage sludge is a by-product generate from wastewater treatment process. The sewage sludge contains significant trace metal such as Cr, Mn, Ni, Cu, Zn, As, Cd and Pb which are toxic to the environment. Sewage sludge is disposed of by landfilling method. However, this option not suitable because of land restriction and environmental control regulations imposed. Therefore, sewage sludge from wastewater treatment plant was incorporated into fired clay brick to produce good quality of brick as well as reducing heavy metals from sludge itself. Sewage sludge with 0%, 1%, 5%, 10% and 20% of were incorporated into fired clay bricks and fired at 1050°C temperature with heating rates of 1°C/min. The brick sample then crushed and sieved through 9.5 mm sieve for Synthetic Precipitation Leaching Procedure (SPLP). From the results, incorporation up to 20% of sewage sludge has leached less heavy metals and compliance with USEPA standard.

1. Introduction

Sewage sludge is a final product from wastewater treatment processes and carried away by the drainage system. Wastewater likewise named as sewage effluent can be perceived as used water originating from domestic, industrial and sewage runoff. Sewage generated from sources point is being conveyed via sewerage systems to treatment facility [1]. According to previous studies, important heavy metals from sewage sludge which can be dangerous to human health includes arsenic, nickel, cadmium, chromium, lead, mercury and nickel [2]. Same studies have found high concentration of heavy metals and radionuclide also gives effect towards human health [3]–[5]. A significant amount of post-consumer sewage sludge waste is directly thrown away into landfills. However, this method may cause many problems towards the environment such as groundwater pollution from landfill leachate, odor emission and soil contamination [6]. Therefore, instead of disposing directly to landfill, recycling and reuse it as a raw material to be incorporated with the clay brick is an alternative disposal method.

*[1] aeslina@uthm.edu.my

Since building materials such as clay brick are produced by using existing natural resources, the feasibility of clay allows its combination with different types of wastes without significant modification of the production process and final product usage [5-6]. Many researchers have successfully attempted the utilization of sewage sludge into fired clay bricks, however, the study on leaching effects still not adequate [7-8]. With rising social awareness about toxic emissions from the sludge and the expanding worry over the disposal of sludge onto landfill, it seems evident that the incorporation of waste as a building and construction raw material can be considered as a vital stride in the correct course [9-10].

Therefore, sewage sludge from wastewater treatment plant was incorporated into fired clay brick to provide an alternative disposal method, while producing adequate brick. The minimization of heavy metal leachability to the environment also was conducted by using Synthetic Precipitation Leaching Procedure (SPLP) [11].

2. Materials and methods

2.1. Material

The sewage sludge waste as in Figure 1 was collected at wastewater treatment plant at Indah Water Konsortium Parit Raja, Johor. Clay soil was obtained at Yong Peng, Johor. Both of the materials, clay soil and sludge were kept properly in a closed container. In this study, the leachate from the brick samples was tested and the results obtained were compared with those from the United State Environmental Protection Agency (USEPA).

Figure 1. Sludge collection from wastewater treatment plant

2.2. Sample Preparation and Heavy Metals Determination

Figure 2 summarizes the sample preparation process and testing in this study. Upon delivery, fresh sewage sludge and clay soil were oven dried separately at 105°C within 24 hours. This process is to remove biological bound water of raw materials and break degradation process of the sewage. The preliminary work was done by analyse sludge using X-Ray Fluorescence (XRF) method to determine heavy metals concentration of the sample. Therefore, the result from XRF have shown that Chromium (Cr), Manganese (Mn), Nickel (Ni), Copper (Cu), Arsenic (As), Zinc (Zn), Lead (Pb) and Cadmium (Cd) are the investigate parameters in this study.

2.3. Brick Manufacturing Process

Trial mix was conducted to determine suitable percentage of sewage sludge content for brick manufacturing. The ratio of sludge was calculated based on the weight of the sludge. Therefore, the manufactured bricks were designed as Control Brick (CB) for brick without sludge waste, SSB1%, SSB5%, SSB10% and SSB20% for brick with 1%, 5%, 10% and 20% of sludge waste, respectively. Brick manufacturing process was started by mixing clay soil with 0% up to 20% of waste with water. The wet mixed sample was then pressed into mould sizes 215 mm x 102.5 mm x 65 mm with pressure of 2500 psi. The prepared brick then were kept at room temperature for 24 hours and followed by another 24 hours for oven drying period. The dried bricks finally were fired in a furnace at 1050°C for 18 hours.

2.4. Leachability Test for Manufactured Brick
Synthetic Precipitation Leaching Procedure (SPLP) was conducted according to Method 1312 [7]. The brick samples were divided into four parts and crushed to obtain representative samples. The SPLP (Method 1312) leaching fluid is simulating an acid rain and prepared by diluting sulphuric and nitric acid (60/40 weight percent mixture) H_2SO_4/HNO_3 into distilled water until the pH was 4.2 ± 0.05. The leaching fluid and samples were put inside the sampling bottle. The bottles were placed in a rotary extractor (Figure 3) at 30 rpm with 22°C to 24°C for 18 ± 2 hour. At the end of the extraction, the samples were filtered through 0.7 µm with glass fibre (Figure 4) and were sent to analyse using inductively coupled plasma mass spectrometry (ICP-MS). The investigate parameters are Chromium (Cr), Manganese (Mn), Nickel (Ni), Copper (Cu), Arsenic (As), Zinc (Zn), Lead (Pb) and Cadmium (Cd).

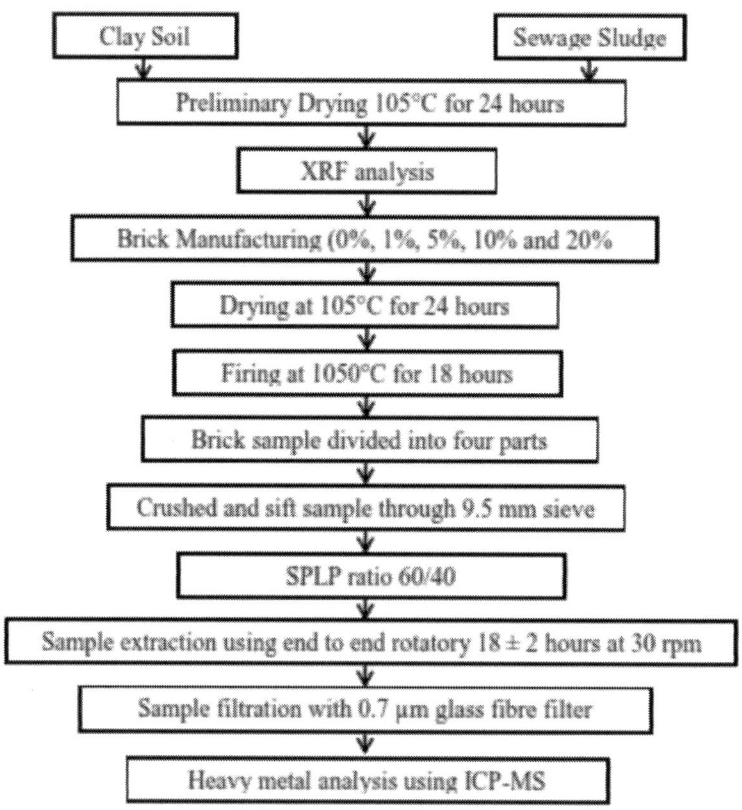

Figure 2. Process used to prepare the samples

Figure 3. Sample extraction using end to end rotatory agitation

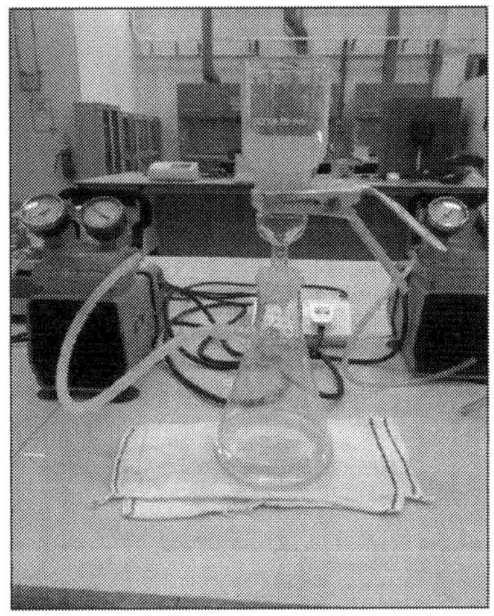

Figure 4. Filtration samples with 0.7 μm glass fibre filter

3. Results and Discussion

3.1. Characteristic of Raw Materials

Table 1 shows the heavy metals concentration using XRF analysis. From the analysis, several parameters show the heavy metal accumulates in the sludge where Mn and Zn are the highest heavy metals with 330 mg/L and 132 mg/L. The lowest heavy metal traced in sludge is Cd with 5 mg/L. Meanwhile, Mn and Cr shows the highest heavy metal in clay soil with 124 mg/L and 100 mg/L. Cd and As are the lowest heavy metals in clay soil. When compares to USEPA, several parameters in both raw materials are over the limitation. Therefore, these parameters are the matter of concern and were selected to be analysing for the next SPLP test.

Table 1. Heavy metals concentration of raw materials using XRF analysis

Heavy Metals	Concentration Limit (mg/l) (USEPA)	Concentration (mg/l)	
		SSW	Clay Soil
Chromium (Cr)	20	49	100
Manganese (Mn)	260	330	124
Nickel(Ni)	8	16	15
Copper(Cu)	800	15	23
Zinc(Zn)	1200	132	33
Arsenic(As)	2.8	55	8
Cadmium (Cd)	0.8	5	0.1
Lead (Pb)	5	56	30

3.2. Leachability for Sewage Sludge Brick

Table 2 shows the leachability result for sewage sludge brick. From the Table 2, Mn and Zn were the highest element traced in Control brick. Meanwhile SSB1%, SSB5%, SSB10% and SSB 20% show the same results where Mn and Zn were the highest element traced in the sample. Others element like Cd and Pb did not leached any or little heavy metals from the brick samples. The results have shown that after incorporation sewage sludge waste into fired clay brick, all the parameters did not exceed USEPA limit. This is due to the phenomenon of heavy metals in sludge is completely vaporised at high firing temperature during sintering process, therefore could reduce the heavy metals inside the brick [7].

Table 2. Heavy metals leachability concentration by using SPLP

Heavy Metals	Concentration Level (mg/l)					USEPA (mg/l)
	Control	SSB1%	SSB5%	SSB10%	SSB20%	
Chromium (Cr)	0.002	0.004	0.004	0.004	0.027	20
Manganese (Mn)	0.518	1.280	1.250	1.220	1.120	260
Nickel (Ni)	0.012	0.004	0.005	0.005	0.005	8
Copper (Cu)	0.006	0.013	0.017	0.015	0.021	800
Zinc (Zn)	0.403	0.681	0.637	0.682	0.689	1200
Arsenic (As)	0.002	0.220	0.378	0.437	0.453	2.8
Cadmium (Cd)	n.d	n.d	n.d	n.d	n.d	0.8
Lead (Pb)	n.d	0.004	0.003	n.d	n.d	5

n.d : Not Detectable

4. Conclusion

As a conclusion, the heavy metals leachability by incorporating different percentages of sewage sludge waste into fired clay brick properties was determined. From the result, concerning on the leachability properties, it can be concluded that incorporation of sewage sludge waste up to 20% into fired clay brick could reduce the heavy metals significantly from the sludge. The results also showed that sewage sludge incorporated into fired clay brick could minimize the toxicity of the heavy metals exist in the raw sludge waste. Therefore, it can be concluded that by incorporating sewage sludge could be an alternative as a replacement raw material for brick manufacturing.

Acknowledgments

The results presented in this paper are part of an ongoing postgraduate research. The authors would like to thank the Faculty of Civil and Environmental Engineering, UTHM for research support.

References

[1] I. Fonts, G. Gea, M. Azuara, J. Ábrego, and J. Arauzo, "Sewage Sludge Pyrolysis for Liquid Production: A Review," *Renewable and Sustainable Energy Reviews*, vol. 16, no. 5. pp. 2781–2805, 2012.

[2] J. A. Cusidó and L. V Cremades, "Environmental Effects of Using Clay Bricks Produced With Sewage Sludge: Leachability and Toxicity Studies," *Waste Manag.*, vol. 32, no. 6, pp. 1202–8, Jun. 2012.

[3] A. N. Victoria, "Characterisation and Performance Evaluation of Water Works Sludge as Bricks Material," *Int. J. Eng. Appl. Sci.*, vol. 3, no. 3, pp. 69–79, 2013.

[4] L. Wang, D. C. W. Tsang, and C.-S. Poon, "Green Remediation and Recycling of Contaminated Sediment by Waste-Incorporated Stabilization/Solidification," *Chemosphere*, vol. 122, pp. 257–64, 2015.

[5] D. Eliche-Quesada, R. Azevedo-Da Cunha, and F. a. Corpas-Iglesias, "Effect of Sludge from Oil Refining Industry or sludge from Pomace Oil Extraction Industry Addition to Clay Ceramics," *Appl. Clay Sci.*, vol. 114, pp. 202–211, Sep. 2015.

[6] C. V. Andreoli and M. Von, *Sludge Treatment and Disposal*, vol. 23, no. 7. 2007.

[7] M. Arsenovic, Z. Radojevic, and S. Stankovic, "Removal of Toxic Metals from Industrial

Sludge by Fixing in Brick Structure," *Constr. Build. Mater.*, vol. 37, pp. 7–14, 2012.

[8] S. N. Monteiro and C. M. F. Vieira, "On the production of fired clay bricks from waste materials: A critical update," *Constr. Build. Mater.*, vol. 68, pp. 599–610, 2014.

[9] A. Abdul Kadir, M. I. H. Hassan, and S. K. H. bin Syed Mohamad, "Leachability of Self-Compacting Concrete (SCC) Incorporated With Fly Ash and Bottom Ash by Using Static Leaching Procedure (SLT)," *Appl. Mech. Mater.*, vol. 773–774, pp. 1261–1265, 2015.

[10] A. A. Kadir and A. Mohajerani, "Physico-Mechanical Properties and Leachate Analysis of Clay Fired Bricks Incorporated with Cigarette Butts," in *International Conference on Environment*, 2008.

[11] Environmental Protection Agency Method 1312, "Synthetic Precipitation Leaching Procedure." pp. 1–30, 1994.

Comparative Study of the Scaling Effect on Pressure Profiles in Capillary Underfill Process

Fei Chong Ng
School of Mechanical Engineering, Engineering Campus, Universiti Sains Malaysia, 14300, Nibong Tebal, Malaysia.
Aizat Abas*
School of Mechanical Engineering, Engineering Campus, Universiti Sains Malaysia, 14300, Nibong Tebal, Malaysia.
M Z Abdullah
School of Mechanical Engineering, Engineering Campus, Universiti Sains Malaysia, 14300, Nibong Tebal, Malaysia.
M H H Ishak
School of Mechanical Engineering, Engineering Campus, Universiti Sains Malaysia, 14300, Nibong Tebal, Malaysia.
Gean Yuen Chong
School of Mechanical Engineering, Engineering Campus, Universiti Sains Malaysia, 14300, Nibong Tebal, Malaysia.
**Corresponding author*: aizatabas@usm.my

Abstract. Optimization of the capillary underfill (CUF) encapsulation process is vital to enhance the package's reliability. Therefore, the design and sizing of the newly developed ball grid array (BGA) device must be considered so that it is compatible with the CUF process. The scaling effect of BGA on CUF flow and its dynamic properties is thoroughly investigated by means of fluid-structure interaction (FSI) numerical simulation. This paper generally highlighted the differences in CUF flow behaviours, together with the pressure distributions between the actual industrial size BGA and the scaled up models for large BGA setup. While flow front profiles appeared to be similar across BGA of various sizes at relative error less than 10%, the CUF filling time gradually increases as the BGA become larger. The scaling limit is found to be at 20, based on the analysis of dimensionless number. The entrant pressure however decreases when the BGA device being scaled up. These findings will assist in the future BGA designs for various sizes used in the CUF encapsulation process.

1. Introduction

Capillary underfill (CUF) encapsulation is important to enhance the electronic package's reliability, cushion it from external contamination, dissipates mechanical streets away from the solder joints as well serves as a heat sink [1]. Hence, in the past decades, various researches were conducted particularly on this field with the aim of improving and further optimize the CUF process.

On the other hand, ball grid array (BGA) is a type of surface mount packaging that utilizes an array of solder bumps as joints. BGA possess various advantages and is introduced to succeed the pin grid array (PGA) that is susceptible to damage due to excessive bending [2]. This generally accounts for the popularity of BGA as the main electronic mounting type used in the industry. Indeed, there have been various studies of CUF encapsulation process. Power-law constitutive equation-based analytical models were constructed to predict the underfilling flow during the encapsulation process [3]. The technique used here is rather cost effective and the CUF flow in the package can be successfully visualized.

For many years, finite volume method (FVM) simulation has been the most favourable method in simulating the CUF flow. Recently, the fluid structure interaction (FSI) library based on FVM simulation has been mainly used in the research works related to encapsulation and electronic packaging, notably in the works by C.Y. Khor [4 – 6]. He demonstrated that FSI simulation is capable of producing highly accurate results rather than relying solely on solving the fluid domain alone. Among his FSI works, various parameters has been analysed: inlet pressure, outlet vent [4], silicon chip thickness [5], solder bump arrangement and sizes [6]. In addition, Lattice Boltzmann method (LBM) has also being adopted to study the encapsulant flow in BGA and similar flow profiles upon side by side comparison to the FVM simulation were obtained [7]. However, the LBM based software is found in certain test cases to be much costly as the computing time is generally longer than using FVM.

Various experimental works have been concerted to validate the numerical simulation study. Most researches generally used a scaled-up flip chip model with replacement fluid instead of the actual device and underfilling machine [3 – 8]. This is mainly to enhance the visualization of the CUF flow across the device and also to minimize the expenditure of the research work. It had been reported that the shape and arrangement of solder bumps in flip chip are seen as the most influential factors in determining the outcome of the CUF process [9]. The CUF filling time of middle empty BGA is found to be shorter compared to the full array BGA and has relatively low number of voids formed [6, 8].

Certain literatures have also investigated the effect of dispensing method on the underfilling encapsulation process. It is reported that L-type dispensing method is regarded as the best option available due to its ability to yield relatively faster and uniform filling [10]. Additionally, air void and incomplete filling are rarely found when L-type dispensing method is being adopted [11]. The presence of void in the encapsulated package will critically affects the reliability of the manufactured package and tend to cause higher stress concentration typically at the proximity of the void region and subsequently leads to device malfunction. Furthermore, the impact of varying the BGA's gap height is investigated and it was reported that the increase in gap height is capable of reducing the void formation as well as further promote the CUF flow [12]. The scaling effect however is not being emphasized here as only one dimension of the BGA gap height is being manipulated, instead of its width and length dimensions.

It appeared that the study of FSI numerical simulation on CUF encapsulation is rather limited in number. Eventually it is also found that there is no research has been conducted to investigate the scaling effect of BGA on the CUF process. To date, there is no existing FSI study on the scaling effect of the CUF encapsulation of BGA device has been found that is concentrated on the pressure and velocity distributions of the CUF encapsulation process. Pressure and its gradient are crucial parameters in deciding the flowability of the encapsulant into the gap of the BGA. Fortunately these parameters is able to be determined precisely through the use of FVM based FSI simulation. Generally this is something that is difficult to be detected through experimental works, unless expensive apparatus are used. The impact of altering the package's size by scaling based on the actual size BGA can be clearly visualized and will provide an insight to package designers on relevant scaling issues. This definitely would aid them in designing future miniature-size or even more complicated device in the future.

2. Problem Descriptions

In Fig. 1, the CUF encapsulation process of BGA flip chip is carried out by dispensing a controlled amount of underfill epoxy mold (EMC) at one side and the encapsulant will flow into the gap between flip-chip and substrate via capillary action. The separation between flip chip and substrate is known as the gap height and is effectively the parameter that will define the strength of the capillary flow [8, 13].

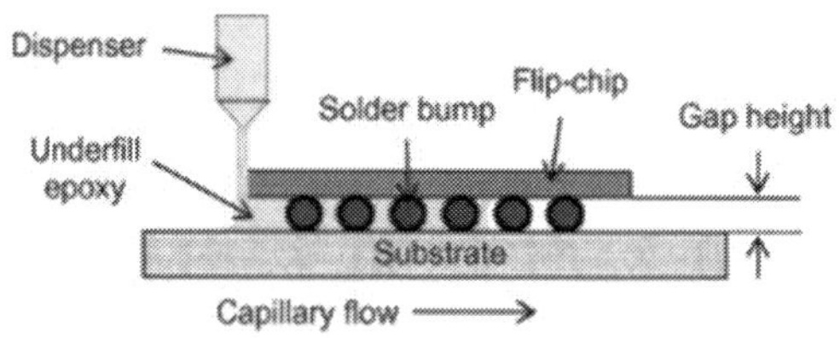

Fig. 1. Schematic of BGA chip underfilling boundary conditions [14].

In this study, a finite volume method (FVM) based numerical simulation is used to visualize the three-dimensional CUF flow in middle empty BGA that is subjected under L-type dispensing method. Middle empty BGA is chosen as the primary subject of research due to its popularity usage in the industrial applications compared to full and perimeter orientations. Similarly, L-type is much more favourable in the industry as the probability of void formation and incomplete filling are relatively less compared to other dispensing method [1]. The combination of L-type dispensing and middle empty orientation for CUF encapsulation is found to be the most ideal setup for an actual industrial application. To account for the effect of different scaling, the CUF process in applied on four scaled-up BGA model of different sizes. Subsequently, all of these models will be compared with an actual miniature size BGA.

3. FSI Numerical Simulations

The numerical simulation of the flow analysis will be based on the well-known Navier-Stokes equation that consists of the continuity, momentum and energy equation that is solved simultaneously during the flow advancement. Furthermore, multiphase formulation is used to distinguish the primary phase (encapsulant) and the secondary phase (air) in CUF flow.

Finite volume method (FVM) based software, Ansys was used to simulate the fluid flows and its dynamic properties in the CUF encapsulation process of middle empty orientation BGA for various scale sizes under L-type dispensing method. Multiphase fluid structure interaction (FSI) was adopted to enhance the accuracy of the simulation works since it will enable precise prediction of the interactions between both fluid and structure domains. The dimensions of the actual size BGA and scaled-up BGA models were shown in Table 1. The material properties of Sn-3.0 Ag-0.5 Cu solder bumps in BGA and the underfill encapsulant used were given in Table 2. The thermal properties of encapsulant were neglected as all CUF dispensing were carried out at same room temperature, such that isothermal is assumed.

Scale	Dimensions of BGA models (mm)			
size	Bump diameter	Bump pitch	Gap height	Length
1.0	0.5	1.0	0.45	12.5
4.4	2.2	4.4	2.0	55.0
6.0	3.0	6.0	2.7	75.0
8.0	4.0	8.0	3.6	100.0
11.0	5.5	11.0	5.0	137.5

Table 1. Dimensions of actual size BGA (scale size 1.0) and various scaled-up BGA models.

Sn-3.0 Ag-0.5 Cu solder bump			Underfill encapsulant		
Material property	Unit	Value	Material property	Unit	Value
Density	g/cm^3	7.04	Density	g/cm^3	1.8
Elastic modulus	GPa	37.4	Molecular weight	kg/kg·mol	44.05
Poisson's ratio	-	0.34	Viscosity	kg/m·s	2.2

Surface tension	N·m	0.06

Table 2. Material properties of the Sn-3.0 Ag-0.5 Cu solder bump and underfill encapsulant used.

Path conforming tetra meshing technique is used in all of the simulation works presented here. This meshing technique was selected to create high quality mesh that are properly discretized that compatible with FLUENT. Figs. 2 (*a*), (*b*) and Figs. 2 (*c*), (*d*) depicted the meshed geometry of the structural and fluid domains respectively.

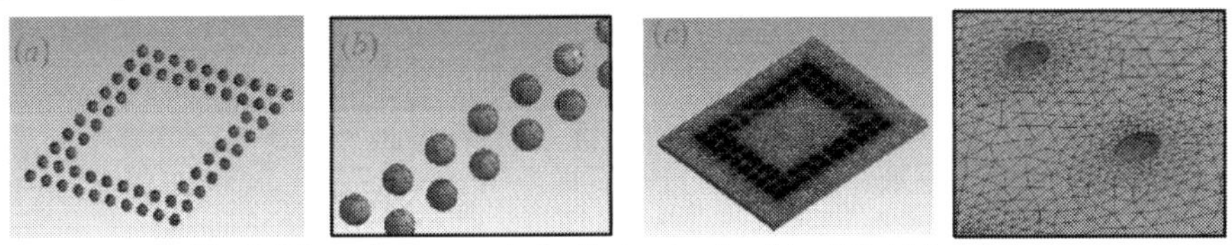

Fig. 2. Meshed geometry in (*a, b*) structural and (*c, d*) fluid domains.

All CUF encapsulation simulations are set under constant temperature, T_0. Both initial inlet and outlet pressures are set to as 0 Pa to demonstrate the capillary flow. Generally, all initial and boundary conditions (BC) are set in the numerical simulation as shown below:

(a) **On wall**: $u = v = w = 0; \quad T = T_0; \quad \frac{\partial p}{\partial n} = 0$

(b) **At inlet**: $P = P_{in}(x, y, z, t); \quad T = T_0$

(c) **On flow front**: $P_{atm} - \frac{\sigma}{R} = P_{atm} - \frac{2\sigma cos\theta}{b}$

L-type dispensing method is modelled in the simulation by designating the sides of the inlet and outlet BC, as illustrated in Fig. 3. A no-slip boundary conditions are applied on all wall surfaces and FSI regions. Wall adhesion setup is enabled to accurately predict the shape of the flow meniscus.

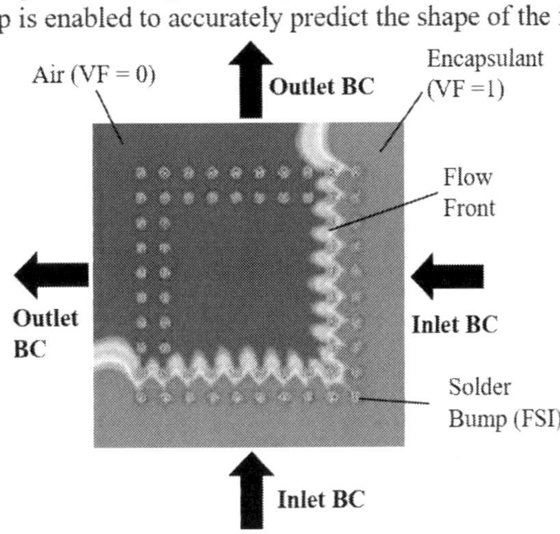

Fig. 3. Boundary conditions illustrated the L-type dispensing method.

FSI simulation is achieved by using the built-in System Coupling module that iteratively coupling the results from Transient Structural and FLUENT. Transient pressure-based solver is selected in the simulation works. Multiphase formulation is enable by specifying the two Eulerian phases that are involved in the simulation models. In the current setup, the encapsulant will be set as the primary phase and air as the secondary phase. Additionally, continuum surface force (CSF) model is applied to simulate the phase interaction between both mentioned phases. Subsequently, an implicit solver is selected as the volume of fluid (VOF) scheme to solve the phase continuity equation in sub time steps together with the momentum and pressure calculations.

4. Results and Discussions

4.1. Validations of Scaled-up BGA Model

To better understand the scaling effect, BGA models of different scale are compared to the actual size BGA based on the front advancement of the encapsulant flow at certain filling percentage. Accordingly, the flow front profiles of each BGA models are compared side-by-side at a selected filling percentage of 40% as shown in Fig. 4. It appears that the flow front profile of CUF flow in the actual size BGA compares well and is found to be almost similar to the profiles in the scaled-up BGA models. Thus, the usage of scaled-up BGA models as a replacement of actual sized device to enhance the visualization aspects in the study of CUF flow is valid as the relative deviation in visualized flow fronts in scaled-up models is within 10%. Furthermore, our validations study does compare well with that reported by Aizat et al. [8] in Fig. 5.

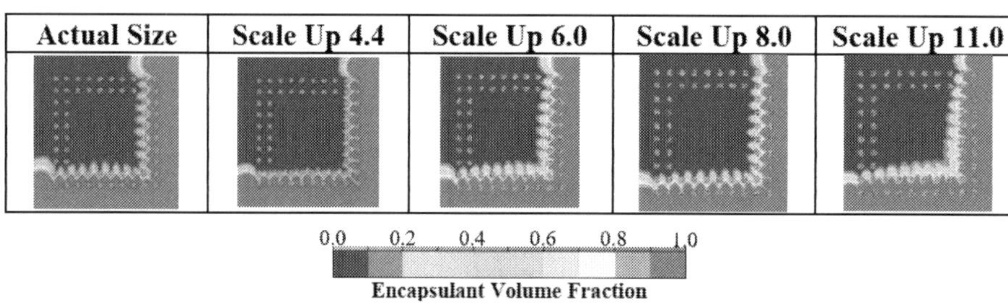

Fig. 4. Comparison of flow front profiles at filling percentage 40% in actual size BGA and scaled-up BGA model.

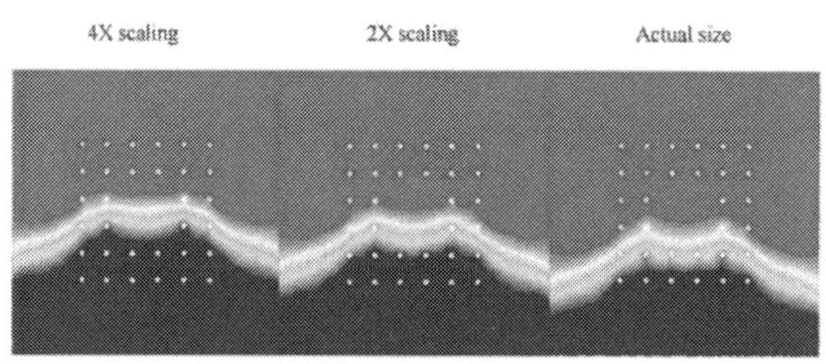

Fig. 5. Comparison of scaling effect and the underfill flow in actual size device based on the works of Aizat et al. [8]

4.2. CUF Filling time of actual size BGA and various scaled-up BGA

The filling time of CUF process at different percentage of flow front advancement filling for actual size and various scaled-up BGA models are presented in Fig. 6. The flow front advancements in CUF encapsulant process of the actual size and scale-up BGA of magnifications 4.4, 6.0 and 8.0 at filling percentages 20% and 40% are shown to quite similar. However, noticeable increase in the scale-up BGA's filling time can be observed after the filling percentage of 60%. The largest BGA of scale size 11.0 is found to have the largest filling time deviation, nearly thrice that of the actual size BGA; while least deviation is recorded in the smallest BGA with 4.4 scale size. It is noteworthy to mention that the overall trends of flow front advancements are homogenous, as the scaling effect will increases the filling time in certain proportional trend. The plot also indicated that all CUF flows exhibit a decelerating trend which the encapsulant tends to slow down toward of the end of the CUF process. This is due to the cumulative resistance of progressive array of solder bumps as the flow progresses.

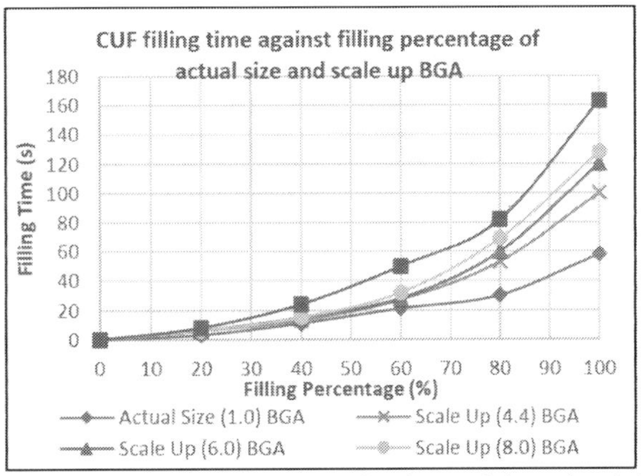

Fig. 6. Filling time against filling percentage of CUF flow in actual size and scaled-up BGA models.

4.3. CUF Pressure Variations near the Dispensing Inlet

Entrant pressure at the dispensing inlet plays a vital role in the CUF process to indicate capability of the encapsulant to fill in the gap of between the cavities via capillary action. The inlet gauge pressures of CUF flow at different filling stages, as indicated by filling percentage, for actual size BGA and scaled-up models are plotted in Fig. 7.

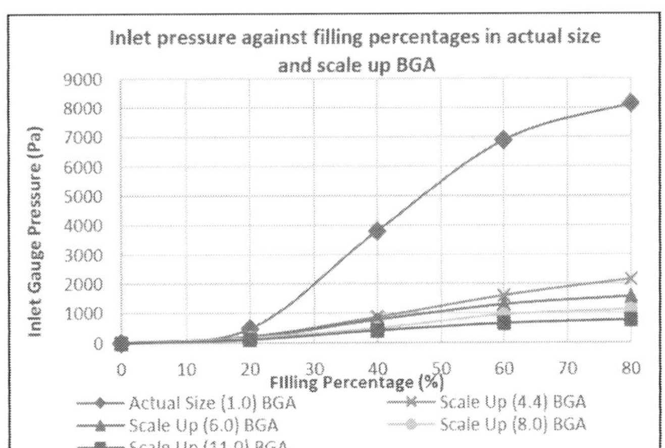

Fig. 7. Inlet gauge pressure at different filling percentages for actual size BGA and scaled-up BGA.

It is shown that the overall entrant pressures at all filling percentages decrease with the increase of BGA's scale size and thus the gap height. The small gap produces larger capillary pressure for the CUF flow and register a larger entrant pressure. It is found that the inlet pressure of the scale up BGA diminished by approximately the same factor as the scale size of the model itself. This serves as the evident of scaling effect on the dispensing pressure of the CUF flow as the BGA is being scaled up while the inlet pressure trend is shown to decline proportionally, based on the scale size itself. For example, the pressure is reduced by approximately 4 times when the size is scaled up with a factor of 4.4. As the flow progresses through the chip cavity, the inlet pressure gradually increases due to the continuous dispensing that build up the hydrostatics and dynamic pressures. Consequently, the pressure of the encapsulant increases to overcome the resistance during the inelastic collisions between the solder bumps array. Therefore, the outlet pressure gradually increases as the cavity is being filled by the encapsulant, with its peak value recorded at 80% filling percentage.

4.4. CUF Pressure Variations near the Dispensing Outlet

The gauge pressures of CUF flow at selected filling percentages are illustrated in Fig. 8. Generally it is found that all the outlet pressures values are significantly low with a value that is slightly above the atmospheric gauge pressure. These non-zero gauge pressures account for the dynamic pressure to displace the air from the outlet vent. Figure 8 also shows that the scaled-up BGA have a relatively lower outlet pressure in contrast to significantly higher pressure shown at the outlet of actual size BGA. In fact, the trend in outlet pressure is similar as the inlet pressure, where it varies inversely proportionally to the scale size. When the BGA is scaled up by 4.4, the outlet pressure reduces approximately by 4, which is consistent in the view of dimensionless number analysis. Furthermore, the outlet pressure also increases as the CUF flow progresses through the solder balls array. However, as the magnitude of outlet pressure is relatively low for all models and at all underfilling stages, its impact on the encapsulant's flow and filling time are minimal and thus negligible.

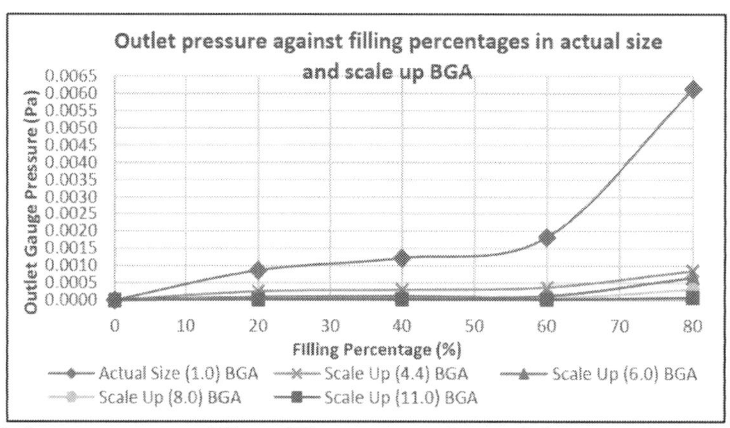

Fig. 8. Outlet gauge pressure at different filling percentages for actual size BGA and scaled-up BGA.

In Fig. 8, the outlet pressure trends in all scaled-up BGA models are similar with a slow increase detected throughout the whole CUF process. Nevertheless, anomalous outlet pressure trend is being observed in the CUF of the actual size BGA model. The pressure increases gradually at the onset of the CUF process before drastically increase at 40% of filling until it attains peak outlet pressure of about 0.006 Pa at 80% of filling. This value is substantially larger than those recorded in the scaled-up BGA models.

5. Conclusions

This paper compares the CUF encapsulant flows in the actual size BGA and different scaled-up BGA models using FVM numerical simulation based on Fluid Structure Interaction (FSI). While the flow front profiles among the actual size BGA and scaled-up BGA models of various sizes are fairly similar and comparable with relative error within 10%, there exists various differences in their CUF process particularly at the filling rate and entrant pressure values. The filling time increases as the size of the BGA becomes larger. In addition, the entrant pressure near the dispensing inlet of the encapsulant also increases as the flow progress and it is inversely proportional to the scale size. The study also recorded that the outlet pressure shares similar trend with the inlet pressure, however its relatively low magnitude can considered negligible in comparison to the overall CUF flow. The numerical data presented in this paper have provided decisive justifications on the impact of scaling effect to the overall flow behaviour of the CUF process of BGA from the perspective of pressure. These findings will provide a good insights for future optimization works on the BGA chip design to attain the optimize CUF encapsulation process as well as overall chip performance and reliability.

Acknowledgements:

This work was partly supported by the FRGS grant FRGS/1/2015/TK03/USM/03/2 and Short Term Grant 60313020 from the Division of Research and Innovation, Universiti Sains Malaysia.

References:

[1] H. Ardebili, M.G. Pecht, "Chapter 3 - Encapsulation Process Technology", Encapsulation Technologies for Electronic Applications, *Elsevier* (2009), pp. 129–179.
[2] J.J. Liu, H. Berg, Y. Wen, S. Mulgaonker, R. Bowlby, A. Mawer, "Plastic ball grid array (PBGA) overview", *Mater. Chem. Phys.* **40** (1995) pp. 236–244.
[3] J.W. Wan, W.J. Zhang, D.J. Bergstrom, "Recent advances in modeling the underfill process in flip-chip packaging", *Microelectronics Journal* (2007), **38**, pp 67–75.
[4] C.Y. Khor, M.Z. Abdullah, F. Che Ani, "Study on the fluid/structure interaction at different inlet pressures in molded packaging", *Microelectronic Engineering* (2011), **88**, pp. 3182 – 3194.
[5] C.Y. Khor, M.Z. Abdullah, "Analysis of fluid/structure interaction: influence of silicon chip thickness in molded packaging, *Microelectronic Reliability* (2013), **53**, pp. 334 – 347.
[6] C.Y. Khor, M.Z. Abdullah, C.S. Lau, W.C. Leong, M.S. Abdul Aziz, "Influence of solder bump arrangements on molded IC encapsulation", *Microelectronic Reliability* (2014), **54**, pp. 796 – 807.
[7] Aizat Abas, MZ Abdullah, MHH Ishak, "Lattice Boltzmann and finite volume simulations of multiphase flow in BGA encapsulation process", *ARPN Journal of Engineering & Applied Sciences* (2015)
[8] Aizat Abas, M.S. Haslinda, M.H.H. Ishak, A.S. Nurfatin, M.Z. Abdullah, F. Che Ani, "Effect of ILU dispensing types for different solder bump arrangements on CUF encapsulation process", *Microelectronic Engineering* (2016)
[9] Y.K. Shen, T.W. Ye, S.L. Chen, C.H. Yin, W.D. Song, "Study on mold flow analysis of flip chip package", *International Communications in Heat and Mass Transfer* (2001), **28**, pp 953-952.
[10] H.R. Gwon, H.J. Lee, J.M. Kim, Y.E. Shin, S.H. Lee, "Dynamic behavior of capillary-driven encapsulation flow characteristics for different injection types in flip chip packaging", *Journal of Mechanical Science and Technology* (2014), **28 (1)**, pp 167-173.
[11] Y.K. Shen, S.T. Huang, C.J. Chen, S. Yu, "Study on flow visualization of flip chip encapsulation process for numerical simulation", *International Communication in Heat and Mass Transfer* (2006), **33**, pp 151-157.
[12] C.Y. Khor, M.Z. Abdullah, F.C. Ani, "Underfill process for two parallel plates and flip chip packaging", *International Communications in Heat and Mass Transfer* (2012), **39**, pp 1205–1212.
[13] Aizat Abas, FC Ng, MZ Abdullah, MHH Ishak, CY Chong, "CUF Scaling Effect on Contact Angle and Threshold Pressure", *to be appeared in Soldering and Surface Mount Technology*
[14] Available at "http://www.epoxysetinc.com/underfill-csp-bga/" (Assessed at 27 September 2016)

IOP Conf. Series: Materials Science and Engineering **203** (2017) 012013 doi:10.1088/1757-899X/203/1/012013

Scaling Effect on Velocity Profiles in Capillary Underfill Flow

Fei Chong Ng
School of Mechanical Engineering, Engineering Campus, Universiti Sains Malaysia, 14300, Nibong Tebal, Malaysia.
Aizat Abas*
School of Mechanical Engineering, Engineering Campus, Universiti Sains Malaysia, 14300, Nibong Tebal, Malaysia.
M Z Abdullah
School of Mechanical Engineering, Engineering Campus, Universiti Sains Malaysia, 14300, Nibong Tebal, Malaysia.
M H H Ishak
School of Mechanical Engineering, Engineering Campus, Universiti Sains Malaysia, 14300, Nibong Tebal, Malaysia.
Gean Yuen Chong
School of Mechanical Engineering, Engineering Campus, Universiti Sains Malaysia, 14300, Nibong Tebal, Malaysia.

**Corresponding author*: aizatabas@usm.my

Abstract. In this paper, the scaling effect of ball grid array (BGA) device on the capillary underfill (CUF) flow and its velocity profiles is thoroughly investigated by means of fluid-structure interaction (FSI) numerical simulation. It is found that generally the flows front profiles attained from device of different scale sizes are comparable to the actual miniature BGA, with relative error approximately under 10%. Based on dimensionless number analysis, the scaling limit is estimated at 20, to maintain low scaling error. The velocity profiles attained on the CUF flow in each models of different scales are relative similar in magnitude and trend. Although the increases in gap height reduces the strength of capillary flow, the flow's velocity still be maintained and the scaling effect is counter-balance with the increases in driving pressure force. The magnitude of entrant velocity is higher at earlier stage of CUF (less than 40% filling); while higher magnitude of exit velocity is found at later stage of CUF (beyond 60% filling). Additionally, the pressure and velocity distributions of CUF flow in miniature device were also studied thoroughly.

1. Introduction

Capillary underfill (CUF) encapsulation is important to enhance the electronic package's reliability, cushion it from external unfavourable contamination, dissipates mechanical streets away from the solder joints as well serves as a heat sink [1, 2]. Hence, various researches were conducted in the past decades on this underfill and electronic packaging field, aimed to substantially improve and also optimize the CUF process. A cost-effective methodology is found in the underfill study, by developing the power-law constitutive equation-based analytical models to predict the underfilling flow [3]. Besides, finite volume method (FVM) simulation has been the most favourable method in simulating the CUF flow. Fluid structure interaction (FSI) scheme based on FVM numerical simulation have been incorporated in the underfill encapsulation study. Using this FSI approach, various underfill parameters has been analysed: inlet pressure, outlet vent [4], silicon chip thickness [5], solder bump arrangement and sizes [6] to optimize the CUF process. Another numerical tool known as Lattice Boltzmann method (LBM) also being adopted to investigate the underfill process. Similar flow profiles upon side by side comparison to the FVM simulation were obtained from LBM simulation [7]. However, the LBM based software is found in certain test cases to be much costly as the computing time is generally longer than using FVM. Furthermore, various experimental study have been conducted along with numerical simulation for validation purpose. Most researches generally used a scaled-up flip chip model with a replacement fluid instead of the actual device, underfill fluid and underfilling machine [3 – 8]. This is mainly to enhance the visualization of the CUF flow across the device and also to significantly minimize the expenditure of the research work.

It had been reported that the geometrical shape and arrangement of solder bumps in flip chip are decisive factor in to the overall CUF process [9]. The CUF filling time of middle empty BGA is found to be shorter compared to the full array BGA and has less air voids occurrence [6, 8]. It is found also that the L-type dispensing method is regarded as the best option available due to its ability to yield relatively faster and uniform filling [10]. Additionally, air void and incomplete filling are rarely found when L-type dispensing method is being adopted [11]. Furthermore, the impact of varying the BGA's gap height is investigated and showed that the increase in gap height is capable of reducing the void formation as well as further promote the CUF flow [12]. The scaling effect however is not being emphasized here as only one dimension of the BGA gap height is being manipulated, instead of its width and length dimensions.

Generally, the study of FSI numerical simulation on CUF encapsulation is rather limited in number, while the scaling effect of BGA on the CUF process is scarcely investigated. To date, there is no existing numerical FSI study on the scaling effect of the CUF encapsulation of BGA device has been found that is concentrated on the velocity distributions of the CUF encapsulation process. The velocity is essentially an important parameter in deciding the flowability of the encapsulant into the gap of the BGA. Extracting velocity data from underfilling fluid is difficult to be achieved through experimental works, since sophiscasted equipment were required. Fortunately, numerical simulation like FVM able to provide data on the velocity profiles. Therefore, through this numerical study, the impact of altering the package's size by scaling on flow's velocity and dynamics can be clearly visualized. This definitely will provide an insight to package designers on relevant scaling issues, to design an optimized underfill process and BGA package.

2. Problem Descriptions

Fig. 1 depicts the capillary underfill (CUF) encapsulation process of BGA flip chip. Controlled amount of underfill epoxy mold (encapsulant) is dispensed at one or more side. Subsequently, the encapsulant will flow into the gap between flip-chip and substrate via capillary action. The separation between flip chip and substrate is known as the gap height. It is the parameter that defines the strength of capillary flow [8, 13].

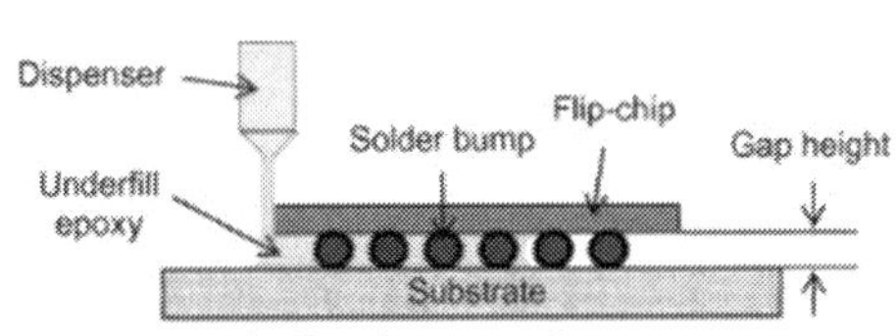

Fig. 1. Schematic of BGA chip underfilling boundary conditions [14].

In this study, the three-dimensional CUF flow in middle empty BGA that is subjected under L-type dispensing method is simulated numerically. Middle empty BGA is chosen due to its popularity usage in the industrial applications compared to full and perimeter orientations. Similarly, the L-type dispensing method is much more favourable in the industry as the occurrence of air void and incomplete filling are less likely compared to the U-type and I-type. Meanwhile, to account for the scaling effect, in this present paper, the CUF process of four different scaled-up BGA models were studied together with actual miniature device.

3. FSI Numerical Simulations

The conventional finite volume method (FVM) based software, Ansys FLUENT was used to simulate the fluid flows and its dynamic properties in the CUF encapsulation process of middle empty orientation BGA for various scale sizes under L-type dispensing method. Multiphase fluid structure interaction (FSI) was adopted to enhance the accuracy of the simulation works since it will enable precise prediction of the interactions between both fluid and structure domains. Governing equations of Navier-stokes will be discretized based on FVM scheme and solved numerically, together with VOF equation. The dimensions of the actual size BGA and scaled-up BGA models were shown in Table 1. The material properties of Sn-3.0 Ag-0.5 Cu solder bumps in BGA and underfill encapsulant are presented in Table 2.

Scale size	Dimensions of BGA models (mm)			
	Bump diameter	Bump pitch	Gap height	Length
1.0	0.5	1.0	0.45	12.5
4.4	2.2	4.4	2.0	55.0
6.0	3.0	6.0	2.7	75.0
8.0	4.0	8.0	3.6	100.0
11.0	5.5	11.0	5.0	137.5

Table 1. Dimensions of actual size BGA (scale size 1.0) and various scaled-up BGA models.

Sn-3.0 Ag-0.5 Cu solder bump			Underfill encapsulant		
Material property	Unit	Value	Material property	Unit	Value
Density	g/cm^3	7.04	Density	g/cm^3	1.8
Elastic modulus	GPa	37.4	Molecular weight	kg/kg·mol	44.05
Poisson's ratio	-	0.34	Viscosity	kg/m·s	2.2
			Surface tension	N·m	0.06

Table 2. Material properties of solder bump and underfill encapsulant.

Path conforming tetra meshing technique is primarily used in all simulation works presented here to create high quality mesh that are properly discretized that compatible with FLUENT. Figs. 2 (*a*), (*b*) and Figs. 2 (*c*), (*d*) depicted the meshed geometry of the structural solid and fluid domains respectively.

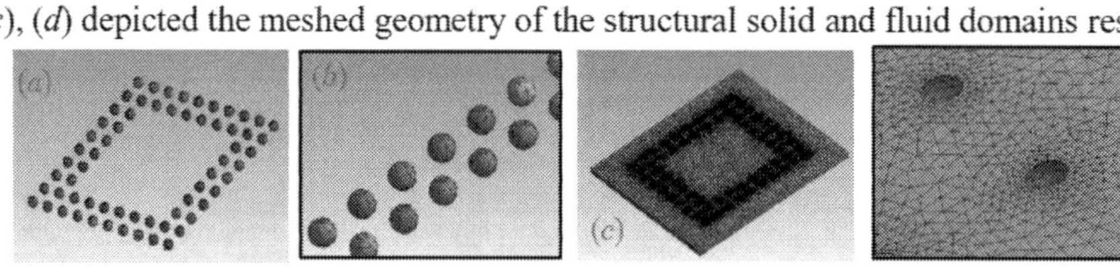

Fig. 2. Meshed geometry in (*a, b*) structural solid and (*c, d*) fluid domains.

All CUF encapsulation numerical simulations are set under constant temperature, T_0. Both initial inlet and outlet gauge pressures are set to as 0 Pa (equal to atmospheric pressure) to model the capillary flow. Generally, all initial and boundary conditions (BC) used are summarized as follow:

(*a*) **On wall**: $u = v = w = 0;\quad T = T_0;\quad \frac{\partial p}{\partial n} = 0$
(*b*) **At inlet**: $P = P_{in}(x, y, z, t);\quad T = T_0$
(*c*) **On flow front**: $P_{atm} - \frac{\sigma}{R} = P_{atm} - \frac{2\sigma\cos\theta}{b}$

L-type dispensing method is modelled in the simulation by designating the sides of the inlet and outlet BC, as illustrated in Fig. 3. No-slip boundary conditions is applied on all wall surfaces and FSI regions. Wall adhesion setup is enabled to accurately predict the shape of the flow meniscus. FSI simulation is achieved by using the built-in System Coupling module that iteratively coupling the results from Transient Structural and FLUENT. Transient pressure-based solver is selected in the simulation works. Multiphase formulation is enable by specifying the two Eulerian phases that are involved in the simulation models where the air will be set as the primary phase while underfill encapsulant as the secondary phase.

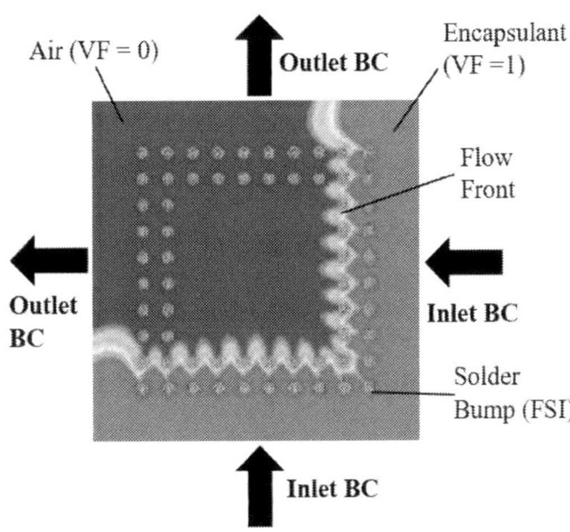

Fig. 3. Boundary conditions illustrated the L-type dispensing method.

4. Results and Discussions

4.1. Flow Front Comparison between Scaled-up BGA Model and Actual Size Device

All the BGA models of different scaling are compared side-by-side to the actual size BGA, in term of front advancement of the encapsulant flow at 40% filling, as shown in Fig. 4. Eventually it appears that the flow front profile of CUF flow in the actual size BGA compares well and is found to be almost similar to the profiles in the scaled-up BGA models, with approximate disparity within 10%. It is showed that scaled-up BGA models can be used to replace the actual miniature device to enhance the visualization aspects in the study of CUF flow. This validation study also compares well with that reported by Aizat et al. [8] in Fig. 5.

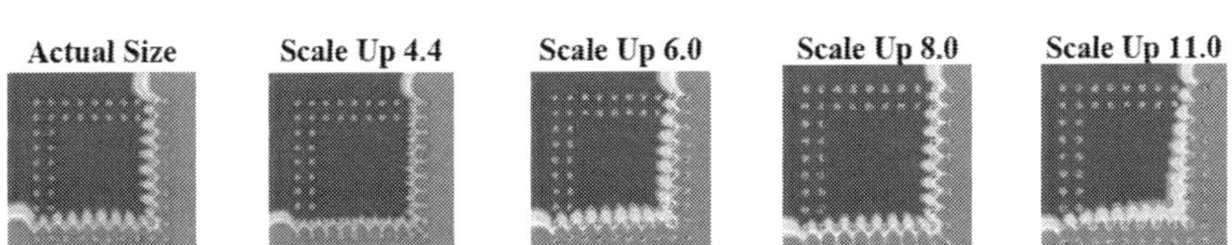

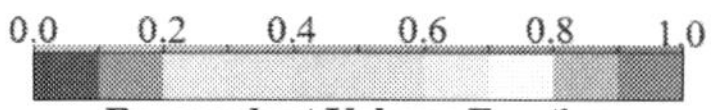

Fig. 4. Comparison of flow front profiles at 40% filling in actual size BGA and scaled-up BGA model.

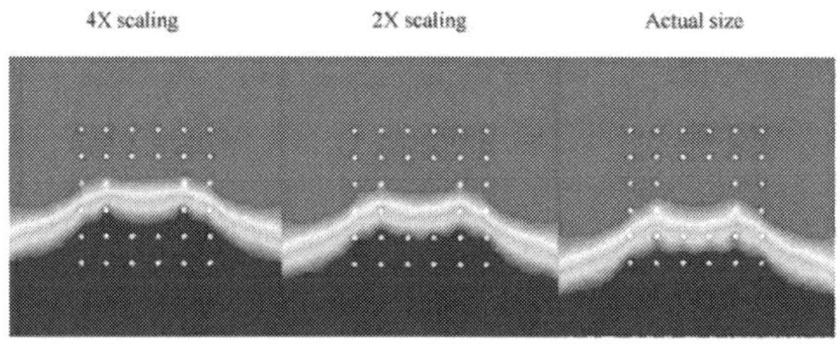

Fig. 5. Comparison of scaling effect and the underfill flow in actual size device based on the works of Aizat et al. [8]

4.2. CUF Velocity Variations near the Dispensing Inlet

In Fig. 6, the velocity of the encapsulant near the inlet for all BGA models shows a qualitatively similarly trend, with its magnitudes falling in the range of 4.0 – 5.9 mm/s, for except the highest flow velocity of 8.7879 mm/s in the actual size BGA during the 20% filling. This eventually shows that the scaling effect did not affect the kinematic of the encapsulant flow across the BGA since its velocity remains invariant. Capillary pressure reduces as the gap height increases in larger scaled up BGA model. However, the reduction in capillary pressure is compensated by the increase in the quantity of dispensed encapsulant. Consequently, this will increase the hydrostatics pressure of the encapsulant due to its weight component to further reduce the filling time. Generally, the encapsulant's velocity reaches peak value at 20% of filling as observed in the actual size BGA and all scaled-up BGA models. This is because in the beginning of CUF process, the encapsulant only flows pass few rows and experiences lowest cumulative bump's resistance. The substantial high net driving pressure is found for the case of fast encapsulant flow particularly at the onset of CUF encapsulation. Subsequently, at CUF after 40% of filling, this net pressure would reduce substantially due to the increasing cumulative solder bump resistance. This will cause the encapsulant to decelerate and having almost similar velocity thereafter until near-completion of the CUF process.

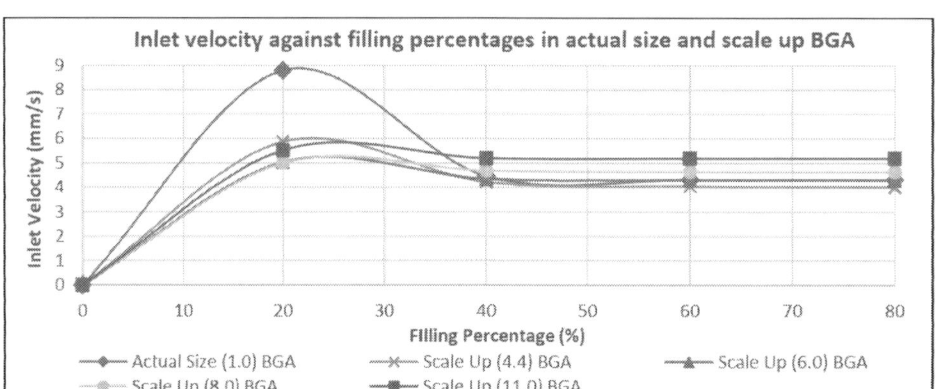

Fig. 6. Inlet velocity of flow at various filling percentages for actual size and scaled-up BGA models.

4.3. CUF Velocity Variations near the Dispensing Outlet

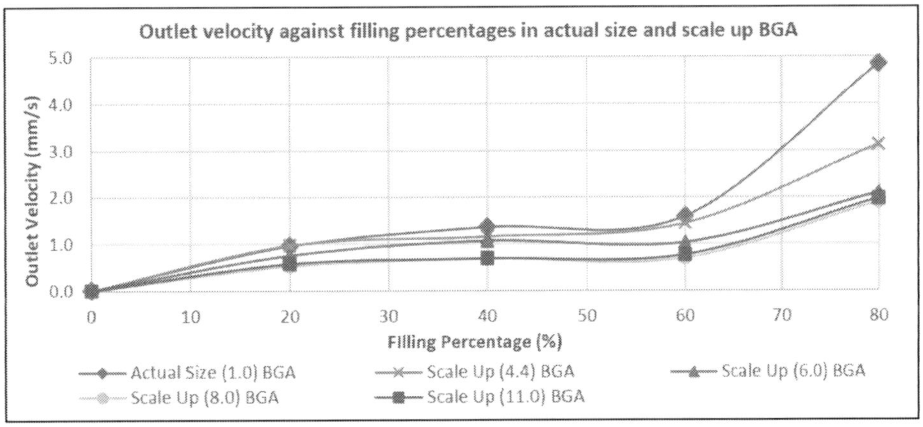

Fig. 7. Outlet velocity of flow at various filling percentages for actual size and scaled-up BGA models.

The velocity of the fluid flow, i.e. the displacing air, near the dispensing outlet vent at different filling percentages for both actual size and scaled-up BGA models are presented graphically in Fig. 7. Upon comparison to the magnitude of encapsulant's inlet velocity, it is found that the rate of air being displaced is slower due to limited quantity of air available in the cavity in contrast to larger quantity of encapsulant in the inlet reservoir that is subjected to the capillary pressure. Moreover, the velocity of the air near the outlet gradually increases as the encapsulant flows during the earlier CUF stages (before 60% of filling) is shown to peak drastically, with highest velocity recorded during 80% of filling. Furthermore, it appeared that the air velocity in actual size BGA is the highest compared to the scaled up BGA at all time. Meanwhile, the 6.0, 8.0 and 11.0 scaled up BGA models have fairly similar outlet vent velocity.

4.4. Pressure and Velocity Distributions in Actual Size Miniature BGA Device

Comparison study conducted previously has revealed that the CUF flows in the actual size BGA is shown to have significantly faster filling rate with high entrant pressure observed. To validate these results, the pressure and velocity distributions of the CUF flow in an actual size BGA are studied. Five specific points, denoted by **P1**, **P2**, **P3**, **P4** and **P5**, are designated along the diagonal region of the PCB, with the points coordinates depicted in Fig. 8. The fluid pressure and velocity variations of the fluid at different filling percentages along these five points are identified and plotted in Fig. 9 and Fig. 10 respectively.

The pressure and velocity distributions obtained in Fig. 8 and Fig. 9 are in good agreement with the results presented earlier. The pressure at **P1** increases as the encapsulant flow progress. Meanwhile the lowest pressure is recorded at **P5** throughout the whole CUF process as it is filled with air. It is noticed that the pressure at particular point will increases rapidly when the encapsulant reached there. On other hand, the fluid's velocity at all points and filling stages are almost similar, with substantially fast fluid flow being observed during 40% at points **P2** and **P3**. This is due to the convergent of flows from both dispensing inlets and the encapsulant that starts to accelerate as it advances toward the bumps-less region. Additionally, the pressure and velocity distributions in the CUF process of actual size BGA is also depicted in Table 3, at 20%, 40%, 60% and 80% fillings.

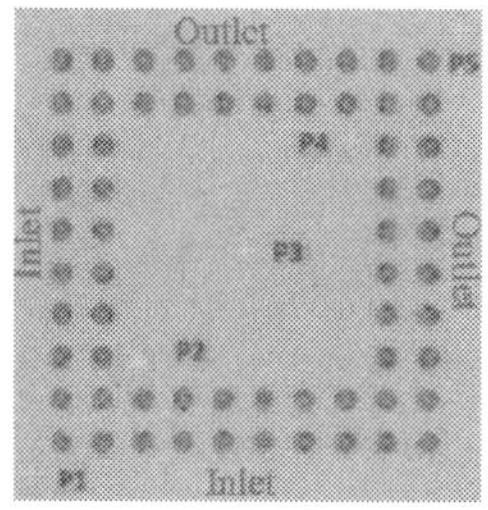

Coordinates (mm)
P1(1.3, 1.3)
P2(3.925, 3.925)
P3(6.55, 6.55)
P4(9.175, 9.175)
P5(11.8, 11.8)

Fig. 8. Locations of the five reference points along the diagonal of actual size BGA device.

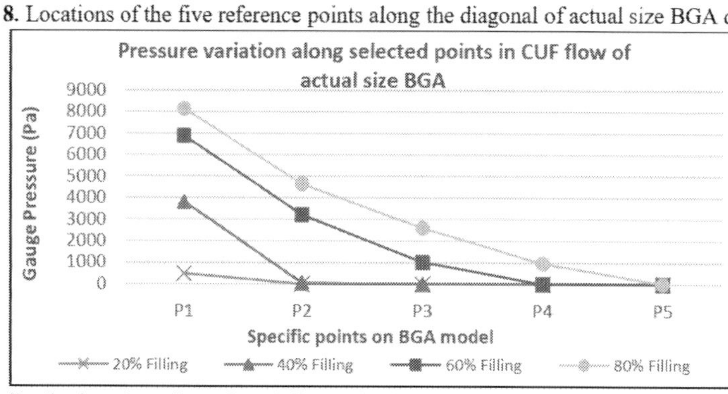

Fig. 9. Pressure distribution along five selected diagonal points in actual size BGA at different filling percentages.

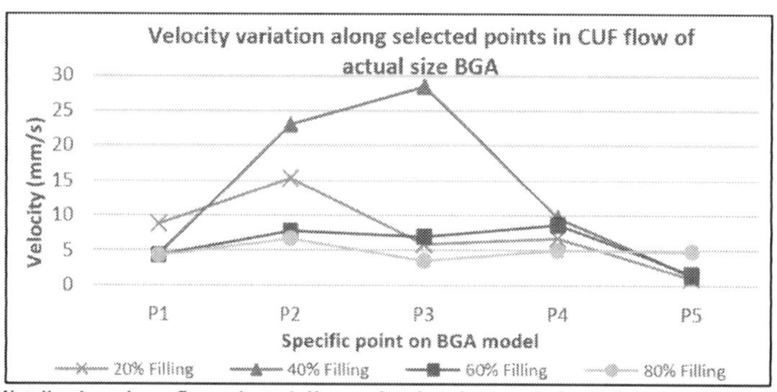

Fig. 10. Velocity distribution along five selected diagonal points in actual size BGA at different filling percentages.

Filling Percentage (%)	20	40	60	80
0 2 4 6 8 10 **Gauge Pressure (kPa)**	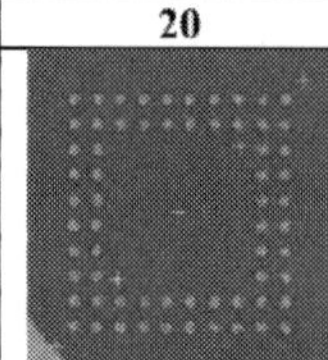	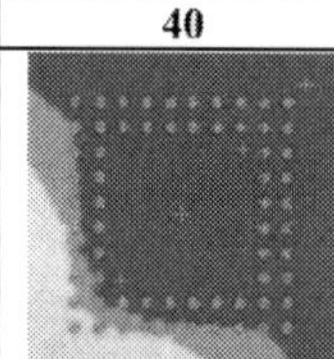	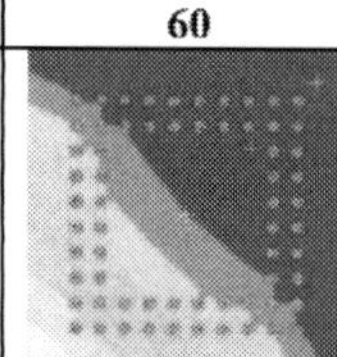	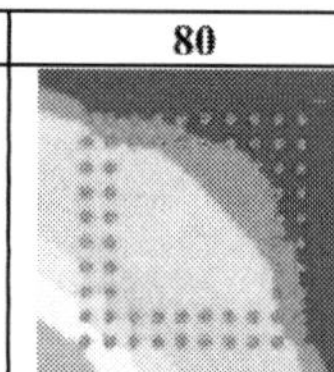
0 6 12 18 24 30 **Velocity (mm/s)**	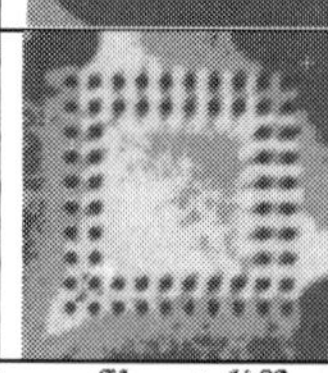	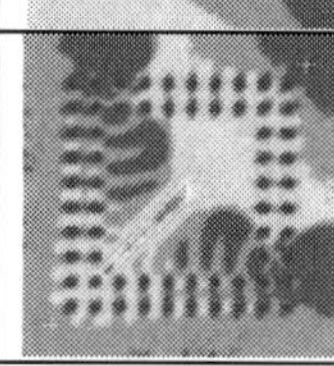	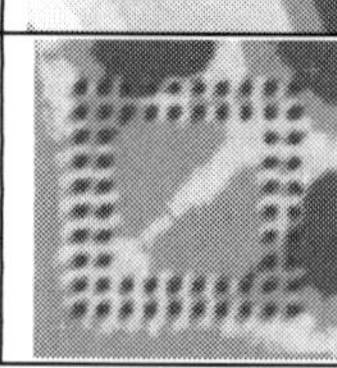	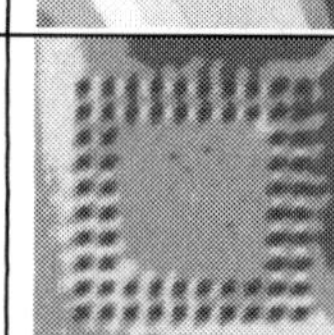

Table 3. Pressure and velocity profiles at different filling percentages for actual miniature BGA.

Generally the pressure profiles in Table 3 show the pressure is gradually building up as the encapsulant flow front progress. The incomplete fill region having zero gauge pressure will indicated that the cavity is full with air at atmospheric pressure. On the contrary, no clear different is observed between the encapsulant-filled region and unfilled (air-rich) region based on the velocity profiles in Table 3. This is because the encapsulant's velocity are almost similar. However, it is noticed that there exists a high velocity region near the diagonal line located at joining points of **P1** and **P2** respectively. This fast flowing encapsulant is due to the joining of flows from both inlets and they eventually start to accelerate due to lesser bump resistance upon entering the bump-less region. Generally the encapsulant flow has a velocity in the range of 0 – 15 mm/s indicating a steady laminar viscous flow. Furthermore, carefully analysis on the velocity profiles eventually found that the encapsulant flows much slower in the surrounding of solder bumps due to kinetic energy loss during the inelastic collision between the encapsulant and the solder bumps. As a result, the encapsulant meniscus velocity decreases gradually until the meniscus detaches from the solder bumps.

5. Conclusions

This paper studied the CUF encapsulant flows in the actual size miniature BGA and various scaled-up BGA models using the Fluid Structure Interaction (FSI) scheme available in commercial FVM software, ANSYS Fluent. Flow front profiles among the actual size BGA and scaled-up BGA models of various sizes are fairly similar and comparable with approximate error not exceeding 10%. This had justified the validity on the usage of scaled-up model in experimental study to enhance visualization aspect. Nevertheless, scaling effect has minimal impact on the fluid's velocity as the CUF flow possess similar inlet and outlet's velocities across all models of various sizes. The inlet's velocity is higher at earlier stage of CUF (less than 40% filling); whereas the outlet's velocity is higher toward the completion of CUF (beyond 60% filling). While the overall encapsulant flows are classified as steady viscous laminar, slower encapsulant speed is being noticed at the vicinity of the solder bumps. The pressure and velocity distributions presented for miniature BGA device illustrated clearly the variations of flow dynamics during the L-type dispensing CUF process. Essentially, these findings will served as a good foundation for future optimization works on the CUF process as well as the design of BGA chip to achieve a desirable chip's performance and reliability.

Acknowledgements:

This work was partly supported by the FRGS grant FRGS/1/2015/TK03/USM/03/2 and Short Term Grant 60313020 from the Division of Research and Innovation, Universiti Sains Malaysia.

References:

[1] H. Ardebili, M.G. Pecht, "Chapter 3 - Encapsulation Process Technology", Encapsulation Technologies for Electronic Applications, *Elsevier* (2009), pp. 129–179.

[2] Ning-Cheng Lee, "9 – BGA and CSP Assembly and Rework", *Reflow Soldering Processes* (2001), pp. 189 – 213.

[3] J.W. Wan, W.J. Zhang, D.J. Bergstrom, "Recent advances in modeling the underfill process in flip-chip packaging", *Microelectronics Journal* (2007), **38**, pp 67–75.

[4] C.Y. Khor, M.Z. Abdullah, F. Che Ani, "Study on the fluid/structure interaction at different inlet pressures in molded packaging", *Microelectronic Engineering* (2011), **88**, pp. 3182 – 3194.

[5] C.Y. Khor, M.Z. Abdullah, "Analysis of fluid/structure interaction: influence of silicon chip thickness in molded packaging, *Microelectronic Reliability* (2013), **53**, pp. 334 – 347.

[6] C.Y. Khor, M.Z. Abdullah, C.S. Lau, W.C. Leong, M.S. Abdul Aziz, "Influence of solder bump arrangements on molded IC encapsulation", *Microelectronic Reliability* (2014), **54**, pp. 796 – 807.

[7] Aizat Abas, MZ Abdullah, MHH Ishak, "Lattice Boltzmann and finite volume simulations of multiphase flow in BGA encapsulation process", *ARPN Journal of Engineering and Applied Sciences* (2015)

[8] Aizat Abas, M.S. Haslinda, M.H.H. Ishak, A.S. Nurfatin, M.Z. Abdullah, F. Che Ani, "Effect of ILU dispensing types for different solder bump arrangements on CUF encapsulation process", *Microelectronic Engineering* (2016)

[9] Y.K. Shen, T.W. Ye, S.L. Chen, C.H. Yin, W.D. Song, "Study on mold flow analysis of flip chip package", *International Communications in Heat and Mass Transfer* (2001), **28**, pp 953-952.

[10] H.R. Gwon, H.J. Lee, J.M. Kim, Y.E. Shin, S.H. Lee, "Dynamic behavior of capillary-driven encapsulation flow characteristics for different injection types in flip chip packaging", *Journal of Mechanical Science and Technology* (2014), **28 (1)**, pp 167-173.

[11] Y.K. Shen, S.T. Huang, C.J. Chen, S. Yu, "Study on flow visualization of flip chip encapsulation process for numerical simulation", *International Communication in Heat and Mass Transfer* (2006), **33**, pp 151-157.

[12] C.Y. Khor, M.Z. Abdullah, F.C. Ani, "Underfill process for two parallel plates and flip chip packaging", *International Communications in Heat and Mass Transfer* (2012), **39**, pp 1205–1212.

[13] Aizat Abas, FC Ng, MZ Abdullah, MHH Ishak, CY Chong, "CUF Scaling Effect on Contact Angle and Threshold Pressure", *to be appeared in Soldering and Surface Mount Technology*

[14] Available at "http://www.epoxysetinc.com/underfill-csp-bga/" (Assessed at 27 September 2016)

Microstructure Analysis of Tungsten Carbide Hardfacing on Carbon Steel Blade

M.Nagentrau [1], A.L.Mohd Tobi [2] Z. Kamdi[1], M. I. Ismail[1], M. Sambu[1]

[1] Faculty of Mechanical and Manufacturing Engineering, Universiti Tun Hussein Onn Malaysia, Batu Pahat, Malaysia

2 Faculty of Engineering Technology, Universiti Tun Hussein Onn Malaysia, Batu Pahat, Malaysia

E-Mail : nagentrau.rau17@yahoo.com

Abstract. Tungsten carbide (WC) hardfacing coating is commonly used to enhance carbon steel blade performance which works in acidic and abrasive condition during production process. This paper deals with tungsten carbide (WC) hardfacing microstructure analysis on a carbon steel blade. Mixing of ilmenite ore with sulphuric acid is performed by the carbon steel blade as part of a production process. Tungsten carbide hardfacing is deposited on the carbon steel blade to enhance its wear resistance. The carbide distribution along with elemental composition analysis of the hardfaced carbon steel blade specimens is examined using Scanning Electron Microscopy (SEM), Energy Dispersive Spectroscopy (EDS) and X-Ray Diffraction (XRD) respectively. Microstructure analysis revealed that different sizes of carbides with non-uniform distribution are found around the coating region. The carbide region is contains high percentage of tungsten (W) meanwhile, non-carbide region rich in tungsten (W) and iron (Fe).

Keywords:

Hardfacing, tungsten carbide, carbon steel blade

1. Introduction

Recently, researchers have shown an increased interest in surface coatings application to improve wear resistance which can prolong life of components [1, 2]. Hardfacing welding approach widely assigned in industries to provide protection for engineering components that was exposed to abrasive environment [3]. The components service life in abrasive environment can be prolonged by hardfacing deposit, especially on the parts that was exposed to production process [4-6]. There is great requirement for high performance material in hardfacing in order to provide protection to the components that operating under corrosive and wear environment [7, 8].

Tungsten carbide (WC) is a suitable hardfacing material that commonly assigned because of good wettability, high hardness and wear resistance [9, 10]. Katsich and Badisch traced the good compatibility of tungsten carbide coatings in protecting engineering components that operates in severe environment [11]. It has been suggested that wear resistance of hardfacing material is dictated by the microstructure characteristics such as carbide distribution. Desai et al. [12] and Shetty et al. [13] revealed that wear property of hardfacing material is highly influenced by amount, size and distribution of carbides in their recent studies. Van Acker et al. highlights the importance of uniform carbide distribution across all the hardfacing coating region in order to improve abrasive wear resistance [14].

Numerous studies have attempted to explain wear characteristics of blades operating in an abrasive condition. Lau et al. investigated wear mechanism of thin edge cutting blade used for grass cutting process using AISI 1090 high carbon steel blades, and concluded wear is mostly because of surface fatigue abrasion in the preliminary stage [15]. Kang et al. performed wear analysis on hardfaced rotary tiller blades and revealed a significant enhancement on protection against wear provided by the hardfacing compared to the un-hardfaced blade [16].

In this study, hardfacing coating microstructure analysis of carbon steel blades operating under severe (acidic and abrasive) environment in a digester tank was studied. In addition, the carbide distribution and elemental composition analyses of a hardfaced CD blade specimen examined under Scanning Electron Microscopy (SEM), Energy Dispersive Spectroscopy (EDS) and X-Ray Diffraction (XRD) respectively.

2. Methodology

2.1 Blade and electrode

The close view of the BS3100 carbon steel Grade A3 blade shown in Figure 1. Carbon steel blade is assigned as the base metal in the hardfacing coating process. In addition, the blade grade is defined by the percentage of carbon in the blade. Carbon steel blade comprise of Carbon (C) and manganese (Mn) as the main element. Additional elements are also present in small portion as presented in Table 1. The hardfacing electrode employed is tungsten carbide (WC) electrode enclosed in tubular mild. High hardness with excellent wear and corrosion resistance promising suitability of tungsten carbide electrode in hardfacing process. The wide usage of tungsten carbide can be seen in hardfacing of cutter knives, construction machineries, pump impellers and speed mullers. The electrode is approximately 6 mm diameter and 400 mm long. Table 2 shows the chemical composition of hardfacing electrode.

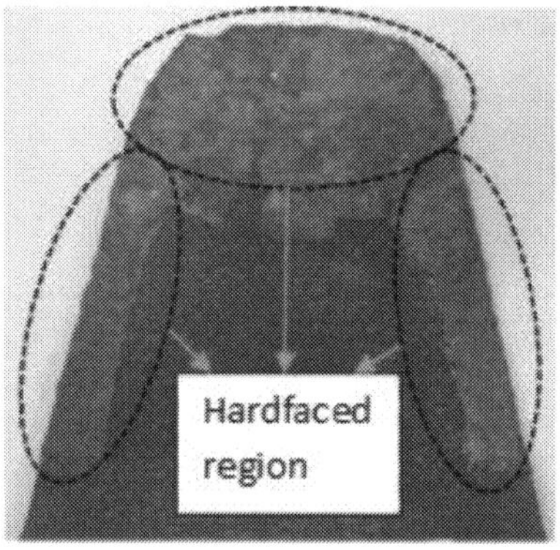

Figure 1: Hardfaced carbon steel blade

Table 1: BS3100 carbon steel chemical composition

Element	C	Mn	Si	S	Mo	P	Ni	Cr
Composition (%)	0.32	0.74	0.25	0.01	0.02	0.02	0.17	0.29

Table 2: Tungsten carbide (WC) electrode Chemical composition

Element	C	Si	Mn	W	Fe
Composition (%)	3.1	0.4	1.5	60.2	balance

2.2. Hardfacing method

One of the fusion/stick welding practise that is used for the purpose of metal combination is Shielded Metal Arc Welding known as SMAW. SMAW welding is considered as economical as no expensive and complex tools are required to perform this type of hardfacing. SMAW hardfacing is performed on carbon steel blade with particular welding condition as presented in Table 3. This welding condition is commonly practiced by the industry for tungsten carbide hardfacing. Figure 2 shows fully hardfaced carbon steel blade that is ready to be fixed in the digester tank for production process.

Table 3: Welding conditions using SMAW welding technique

Welding current	Welding speed	Polarity	Electrode feed rate	Welding width	Electrode length
150 A	0.21-0.24 cm/s	DC	0.34-0.36 cm/s	2 cm	35 cm

(a) (b)

Figure 2: Hardfacing coverage of carbon steel blade using SMAW: (a) top view and (b) front view.

2.3. Specimen preparation

Figure 3 illustrates the sectioning of seven specimens at several positions from the tip of carbon steel blade. Water jet cutting is utilised for specimen sectioning with minimum cutting depth of 2 cm. The surface of the specimens are flattened using table grinder, grit papers and subsequently polishing process as shown in Figure 3. SEM, EDS and XRD analyses are performed on the prepared specimen.

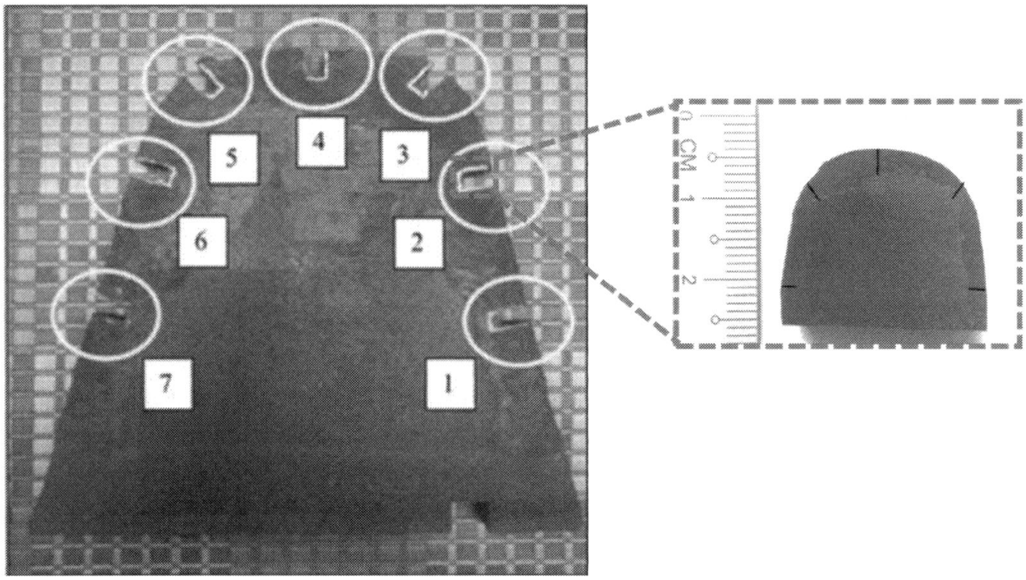

Figure 3: Cross section view of the specimen from carbon steel blade

3. Results and discussion

3.1. Microstructure and elemental composition analyses

SEM is used to analyse the microstructure of the specimens. The backscattered SEM images distinguish the coating and substrate regions by brightness contrast. The distribution pattern of carbide is observed in SEM images. Figure 4 illustrates a close-up view of carbide distribution in the coating. The hardfacing coating region is composed of different sizes of carbide. It is found that large carbides tended to concentrate in the lower part of coating region (near the coating-substrate interface), at about 1 mm in size. The top part of the coating tended toward smaller carbides. Since large carbides are harder and denser, it was deposited at bottom (sunk) of non-carbide metal matrix region and formed at the interface between the substrate and coating. Figure 5 presents close-up view of non-carbide region under SEM. It is noted that the smaller carbide in non-carbide region are not fully grown carbides which are in small grain size and needle like shape. It is believed that the microstructure characteristic such as the type, amount, size, and the distribution of carbides has a great influence on the wear properties. The substrate region is distinguished as being darker compared to the coating region due to the absence of tungsten carbide particles.

IOP Conf. Series: Materials Science and Engineering **203** (2017) 012014 doi:10.1088/1757-899X/203/1/012014

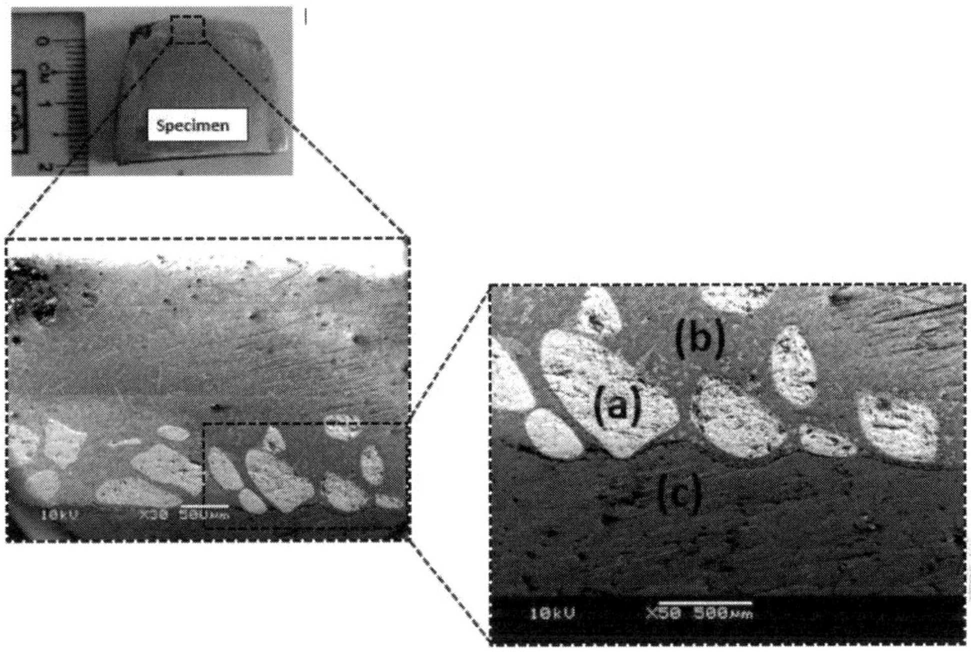

Figure 4: Close-up view of carbide distribution under SEM: a) carbide region, b) non-carbide region, c) substrate

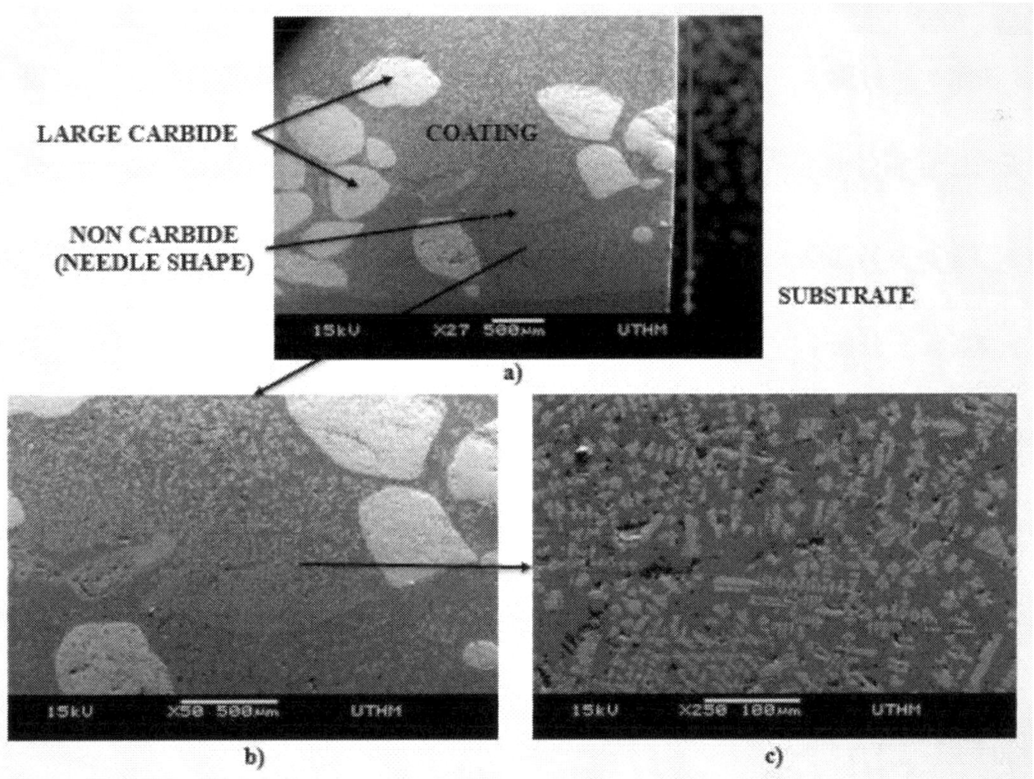

Figure 5: Close-up view of non-carbide region under SEM

IOP Conf. Series: Materials Science and Engineering **203** (2017) 012014 doi:10.1088/1757-899X/203/1/012014

EDS analysis is conducted to determine the elemental composition of the substrate and the coating. EDS analysis results are presented in Figure 6a, 6b and 6c define the carbide region, non-carbide region and substrate region, respectively, based on Figure 4. It is apparent from the figures that tungsten (W), carbon (C), oxygen (O) and iron (Fe) are identified in the EDS analysis. As expected, a high percentage of W is found in the carbide region. The non-carbide region is rich in both W and Fe indicative of carbide and binder in close proximity. The substrate region mainly consisted of Fe. In addition, other elements such as C and O are noted in minimum composition in all regions. Furthermore, the presence of O in both coating and substrate regions is believed to be due to the presence of an oxide layer that formed during the casting process of the blade.

(a)

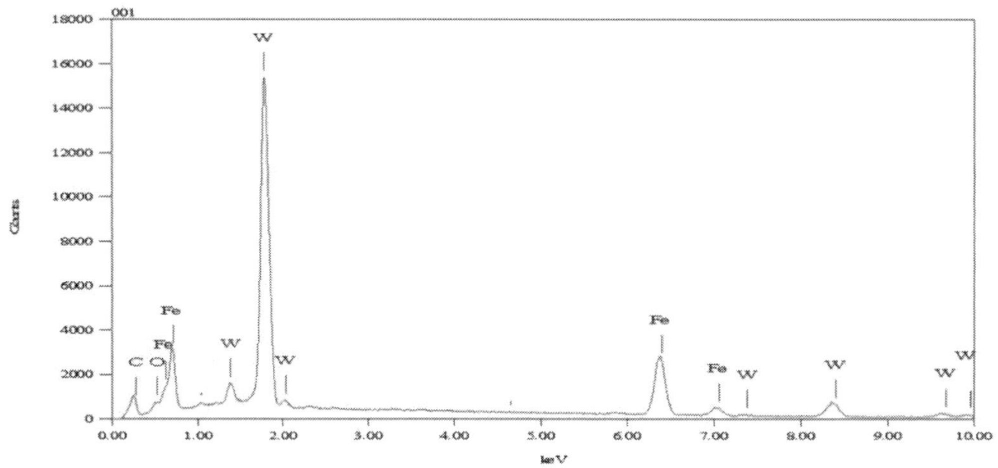

(b)

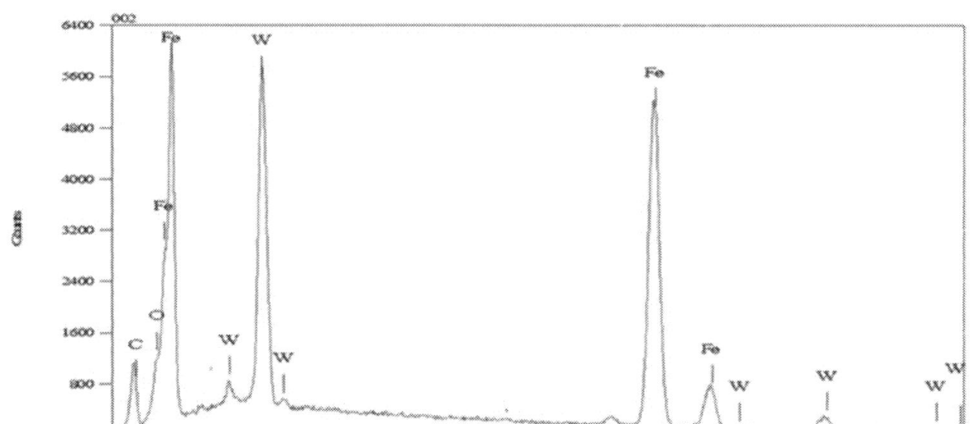

c)

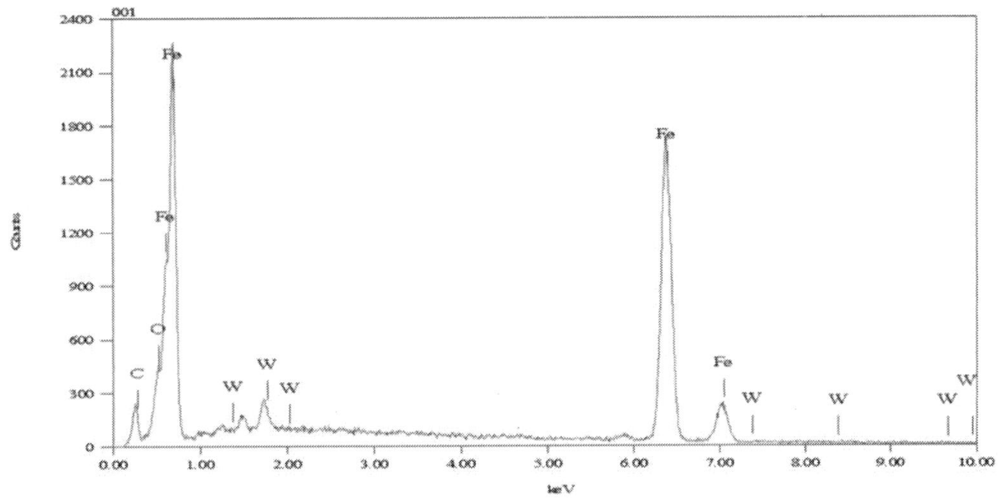

Figure 6: EDS elemental composition analysis of (a) carbide region, (b) non-carbide region and (c) substrate region

Table 4 shows the percentage of the component detect in the XRD analysis for coating. The results of XRD analysis confirmed the presence of Iron (Fe), Tungsten Carbide (W_2C) and Iron Tungsten Carbide (Fe_3W_3C). During hardfacing, the melted Tungsten Carbide (WC) electrode is distributed among iron and as the result of solidification, the iron tungsten carbide was formed. The presence of Fe_3W_3C in these coatings are expected to be of high wear resistance. It is noted that tungsten carbide (W_2C) act as reinforcement particle meanwhile iron (Fe) and iron tungsten carbide (FeW_3C) are the metal matrix in the coating region.

Table 4: XRD analysis on coating

Element/Compound	Percentage (%)
Iron (Fe)	37.8
Tungsten Carbide (W_2C)	20.7
Iron Tungsten Carbide (FeW_3C)	41.4

4. Conclusion

A microstructure analysis of tungsten carbide (WC) hardfacing on a carbon steel blade is studied comprehensively. Microstructure and elemental composition analyses are performed. The following conclusions can be drawn from the obtained results:

- SEM images of the hardfacing showed in detail how the coating was made up of carbide and non-carbide regions. Different sizes of carbides with non-uniform distribution are found around the coating region.
- EDS analysis confirmed high percentage of W is found in the carbide region, meanwhile non-carbide region contains both W and Fe indicative of carbide and binder in close proximity.
- XRD analysis proved the formation of iron tungsten carbide is due to distribution of melted WC electrode among iron the result of solidification.

Acknowledgement

The authors acknowledge the financial support by the Ministry of Education Malaysia and Universiti Tun Hussein Onn Malaysia. This research is supported by the Knowledge Transfer Programme (KTP), Vot. No. 1483.

References

[1] Tobi, A. M., Ding, J., Pearson, S., Leen, S. B., & Shipway, P. H. (2010). The effect of gross sliding fretting wear on stress distributions in thin W-DLC coating systems. Tribology International, 43(10), 1917-1932.
[2] Tobi, A. M., Shipway, P. H., & Leen, S. B. (2011). Gross slip fretting wear performance of a layered thin W-DLC coating: Damage mechanisms and life modelling. Wear, 271(9), 1572-1584.
[3] Hornung, J., Zikin, A., Pichelbauer, K., Kalin, M., & Kirchgaßner, M. (2013). Influence of cooling speed on the microstructure and wear behaviour of hypereutectic Fe–Cr–C hardfacings. Materials Science and Engineering: A, 576, 243-251.
[4] Kim, C. K., Lee, S., Jung, J. Y., & Ahn, S. (2003). Effects of complex carbide fraction on high-temperature wear properties of hardfacing alloys reinforced with complex carbides. Materials Science and Engineering: A, 349(1), 1-11.
[5] Gregory, E. N. (1980). Selection of materials for hardfacing. The Welding Institute, Weld Surfacing and Hardfacing, 22-27.
[6] Güleng, B., & Kahraman, N. (2003). Wear behaviour of bulldozer rollers welded using a submerged arc welding process. Materials & design, 24(7), 537-542.
[7] Marimuthu, K., & Murugan, N. (2013). Prediction and optimisation of weld bead geometry of plasma transferred arc hardfaced valve seat rings. Surface Engineering, 19(2), 143-149.
[8] Deng, H., Shi, H., & Tsuruoka, S. (2010). Influence of coating thickness and temperature on mechanical properties of steel deposited with Co-based alloy hardfacing coating. Surface and Coatings Technology, 204(23), 3927-3934.
[9] Ahmad, N. A., Kamdi, Z., Mohamad, Z., Omar, A. S., Latif, N. A., & Tobi, A. M. (2017). Characterization of WC-10Ni HVOF Coating for Carbon Steel Blade. In IOP Conference Series: Materials Science and Engineering (Vol. 165, No. 1, p. 012022). IOP Publishing.
[10] Tobi, A. M., Kamdi, Z., Ismail, M. I., Nagentrau, M., Roslan, L. N. H., Mohamad, Z. & Latif, N. A. (2017). Abrasive Wear Failure Analysis of Tungsten Carbide Hard facing on Carbon Steel Blade. In IOP Conference Series: Materials Science and Engineering (Vol. 165, No. 1, p. 012020). IOP Publishing
[11] Katsich, C., & Badisch, E. (2011). Effect of carbide degradation in a Ni-based hardfacing under abrasive and combined impact/abrasive conditions. Surface and Coatings Technology, 206(6), 1062-1068.
[12] Desai, V. M., Rao, C. M., Kosel, T. H., & Fiore, N. F. (1984). Effect of carbide size on the abrasion of cobalt-base powder metallurgy alloys. Wear, 94(1), 89-101.
[13] Shetty, H. R., Kosel, T. H., & Fiore, N. F. (1982). A study of abrasive wear mechanisms using diamond and alumina scratch tests. Wear, 80(3), 347-376.
[14] Van Acker, K., Vanhoyweghen, D., Persoons, R., & Vangrunderbeek, J. (2005). Influence of tungsten carbide particle size and distribution on the wear resistance of laser clad WC/Ni coatings. Wear, 258(1), 194-202.
[15] Lau, K. H., Mei, D., Yeung, C. F., & Man, H. C. (2000). Wear characteristics and mechanisms of a thin edge cutting blade. Journal of Materials Processing Technology, 102(1), 203-207.
[16] Kang, A. S., Cheema, G. S., & Singla, S. (2014). Wear Behaviour of Hardfacings on Rotary Tiller Blades. Procedia Engineering, 97, 1442-1451.

Solving Cracking Phenomenon in Premium Transparent Toilet Soap Production Using Stretched LLDPE Film Wrap

Zainon Binti Mat Sharif[1], Norhasnina Binti Mohd Taib[2], Mohd Sallehuddin Bin Yusof[3], Mohammad Zulafif Bin Rahim[4], Abdul Latif Bin Mohd Tobi[5], Mohd Syafiq Bin Othman[6]

[1,2]Graduate Business School, College of Graduate Studies, Universiti Tenaga Nasional, Putrajaya Campus, Jalan IKRAM-UNITEN, Kajang, Selangor, Malaysia.

[3,4,5]Department of Manufacturing And Industrial Engineering, Universiti Tun Hussein Onn, Parit Raja, 86400 Batu Pahat, Johor, Malaysia

[6]Faculty Hotel and Tourism, Universiti Teknologi Mara, Bertam Campus, 13200 Pulau Pinang, Penang, Malaysia.

zainon@uniten.edu.my, norhasninamohdtaib@yahoo.com, mdsalleh@uthm.edu.my, zulafif@uthm.edu.my, abdlatif@uthm.edu.my, mohdsyafiqq@gmail.com

Abstract. Cracking phenomenon in soap production is an imminent problem. It renders the soap aesthetically unpleasing. This study attempts to find best solution to solve cracking phenomenon in premium soap production. The adopted approach is a stamping method with stretched LLDPE wrap film. The result shows that stretched LLDPE wrap film able to solve the cracking problem. The appearance of the premium transparent was improved. This paper presents the results and the SOP for stretched LLDPE film wrap for soap making industries to adopt.

Keywords: cracking phenomenon, toilet soap, LLDPE wrap, stamping method

1. Introduction

Scientifically, soap is a blend of sodium or potassium salts with various naturally occurring fatty acids. Soap are chiefly utilized as surfactants for washing, showering and cleaning. They are additionally utilized as a part of textile spinning and also are important components of lubricants. There are many types of soaps like opaque soap (example brand Lux and May), transparent soap, liquid soap and others. Premium transparent toilet soap is a clear soap with a high glycerine content hence it also referred as glycerine soap. It is recognizably different from other soaps because it is transparent. The clarity is due to the arrangement of the soap molecules, which can be excited through the expansion of alcohol and sugar [1]. The making of premium transparent toilet soap involved lots of processes and techniques. Correct techniques will yield beautiful premium transparent toilet soap and vice versa. One of the problems facing soap production is the cracking phenomenon in premium transparent toilet soap. A cracking phenomenon in the premium transparent toilet soap will give bad expectation to customers. Example of cracking phenomenon in the premium transparent toilet soap is shown in *Figure 1*.

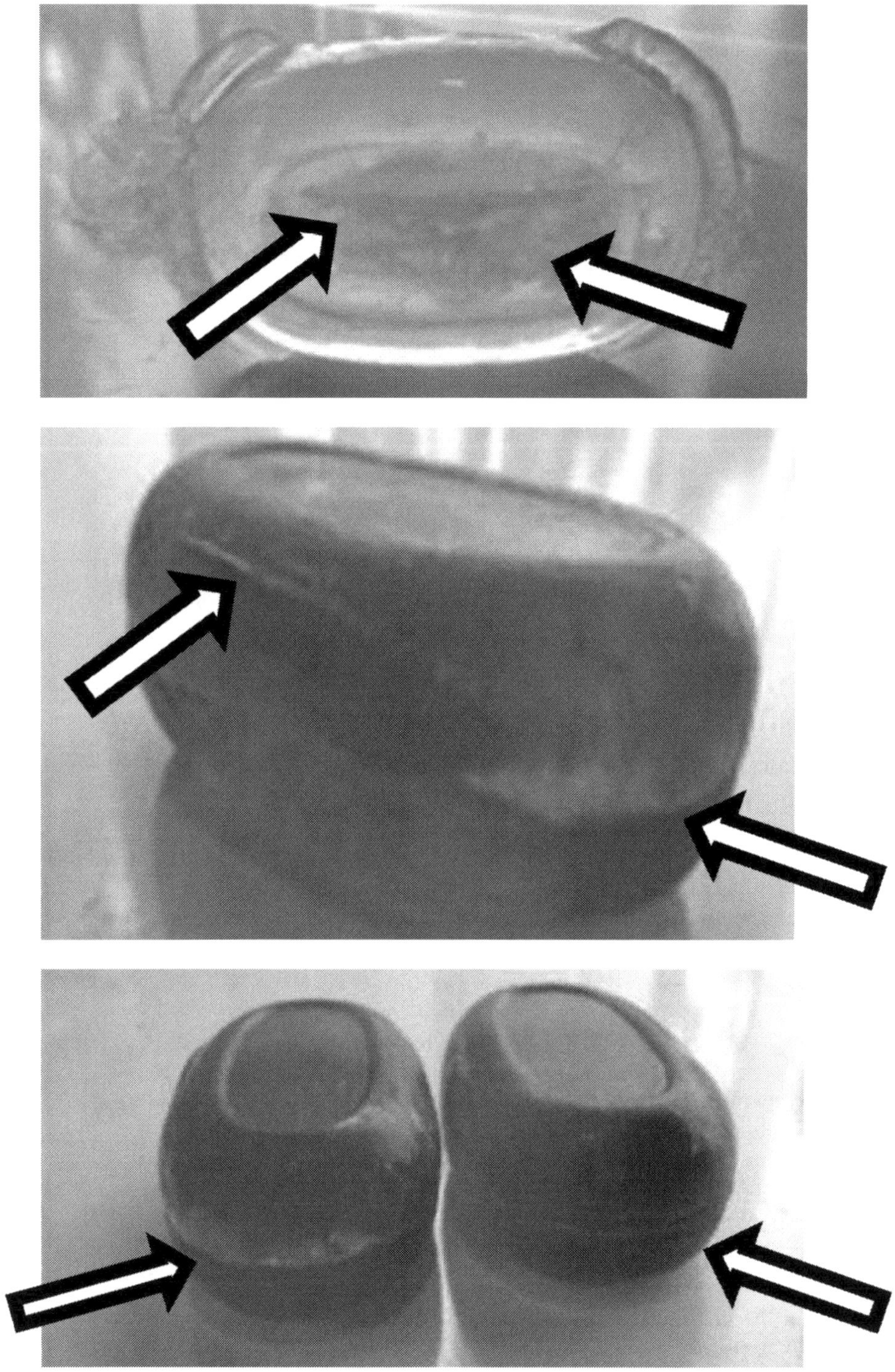

Figure 1: Cracking phenomenon in transparent soap (arrows shown)

The customers claimed that the premium transparent toilet soap that exhibits cracking problem is a rejected soap. Due to the rejection problems, the production cost has increased dramatically and time has been wasted to reproduce the premium transparent toilet soap.

Cracking can be due to a various reasons such as bar design, degree of distortion of the blank during stamping, composition and level of fragrance and other additives, and the efficiency of soap line finishing. There are two types of cracking; wet and dry crack. Wet cracking caused by washing utilization and normally it give rise to ugly lines on the soap bar. It is important to solve this problem in order to ensure quality of the final product. Dry cracks is due to fissure formation via air entrapped in the soap during final extrusion [4].

This paper describes the solution to cracking phenomenon using stamping method with LLDPE wrap film to solve cracking phenomenon. The stretched wrap film material is made from linear low density polyethylene (LLDPE). The function of this stretchable films is to ensure moisture protection and occur during storage. The premium transparent soap will be coated with stretch wrap film before stamping process.

2. Method and Materials

2.1 Preparation of premium transparent toilet soap

Lye solution, oil blend, solvent and sugar solution were prepared according to the following steps (as shown in 2.3 Procedure Flowchart). Oil blend was heated using water bath technique until the temperature of the water reaches 57°C - 62°C. Then, the lye solution was added into the pot and stirred until the thin emulsion forms. After that, solvent was poured into the pot, stirred and wrapped with plastic wrap for 2 hours. After 2 hours, sugar solution was added and allows the base mixed for a few minutes. Lastly, pour the transparent soap base into the mould and allows the base to set up. The transparent soap base was stored for 2 weeks before re-melting and pouring process. The storage was important for stabilizing the chemical reaction in transparent soap base. After 2 weeks, the transparent soap base was re-melted and added with colour and fragrance.

2.2 Preparing the stamping method with stretched wrap film

All the premium transparent toilet soaps were coated with stretch wrap film before stamping process. The temperature of premium transparent toilet soap is 16°C during the stamping process. The pressure of stamping machine used is 8 tonnes.

IOP Conf. Series: Materials Science and Engineering **203** (2017) 012015 doi:10.1088/1757-899X/203/1/012015

2.3 Procedure Flowchart

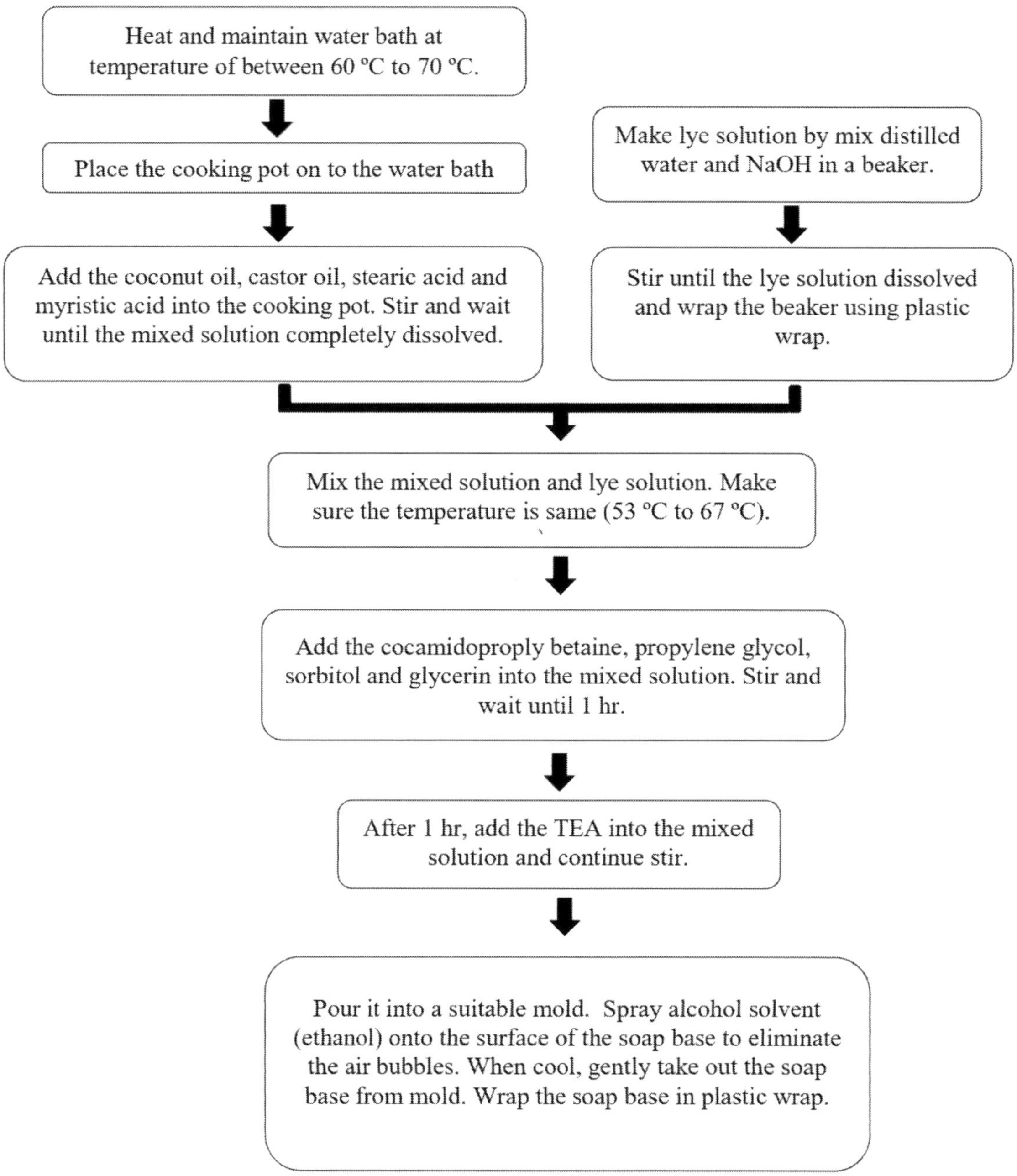

3. Results and Discussion

3.1 Observation of stretched wrap film premium transparent soap.

The results of the stamping method with stretch wrap film are presented in the Figure 2.

Figure 2: Premium transparent toilet soap after stamping method with stretch wrap

Figure 2 shows that there is no cracking problem occurred after stamping method with stretched wrap film. The surface of the premium transparent toilet soap was smooth. This is due to the containment of moisture content in the premium transparent toilet soap. The moisture in the transparent soap will decrease if the transparent soap was exposed to the open air before stamping process. To prevent the moisture loss, transparent soap was wrapped with stretched wrap film. A good transparent soap should have moisture content between 12% and 20% [2]. The function of stretchable films is moisture protection and preventing cracking during storage.

Stretched wrap film is a highly stretchable plastic film that can wrapped around the items. The elastic recovery keeps the items tightly bound. Stretch wrap that was used in this study is shown in Figure 3.

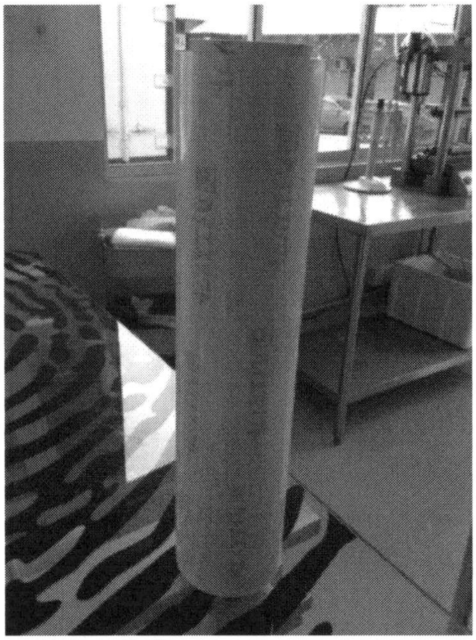

Figure 3: Stretch wrap

The material for the stretch wrap film is low density polyethylene (LLDPE). Granules of linear low density polyethylene are shown in *Figure 4*.

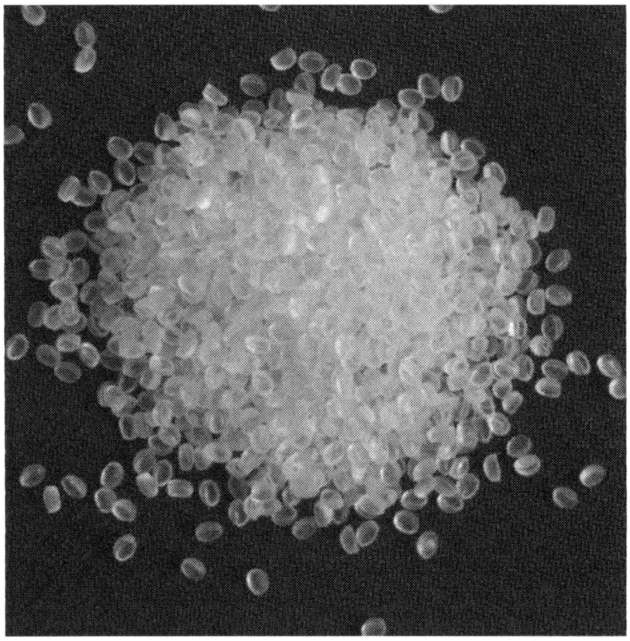

Figure 4: Granules of linear low density polyethylene

LLDPE is a considerably linear polymer (polyethylene) with significant quantities of short branches commonly made by copolymerization of ethylene with long chain olefins (Wikipedia, 2015). The properties of stretched wrap film such as strength and clarity. It makes the transparent soap more clear after stamping method. The benefit of the LLDPE stretch wrap film are ability to protect the transparent soap from exposure to the elements, speed up movement went wrapping process, excellent visual characteristics for better presentation and easy to use.

4. Conclusion

This study shows that stretched wrap film can solve cracking phenomenon problem as evidence by the smooth soap production. Soap making industry can adopt and adapt this procedures to remove cracking problem.

Acknowledgement

The authors would like to express greatest appreciation to the Dean of College of Graduate Studies (COGS), Universiti Tenaga Nasional (UNITEN) for giving the permission to publish this paper.

References

[1] Grosso, A. (2007). The Everything Soapmaking Book: Recipes and Techniques for Creating Colorful and Fragrant Soaps. 2 ed. Adam Media. ISBN 1-59869-229-1.
[2] Kuntom, A. and Roddy, M.R. (2009). Palm-Based Transparent Soap Slab. MPOB information series. No.433. ISBN 1511-7871.
[3] Linear-Low Density Polyethylene. (2015). Retrieved on 29/6/2015 from https://en.wkipedia.og/wiki/Linear_low-density_polyethylene/
[4] Wood, T.E. (1996). Soaps and Detergents – A Theoretical and Practical Review, edited by Luis Spitz, AOCS Press, Champaign,p.46.

IOP Conf. Series: Materials Science and Engineering **203** (2017) 012016 doi:10.1088/1757-899X/203/1/012016

Optimization of bone drilling parameters using Taguchi method based on finite element analysis

Ayip Rosidi[1], Turnad Lenggo Ginta[1], Ahmad Majdi Bin Abdul Rani[1]

[1]Mechanical Engineering Department, Universiti Teknologi PETRONAS, Perak 31750, Malaysia

E-mail: turnad.ginta@utp.edu.my

Abstract. Thermal necrosis results fracture problems and implant failure if temperature exceeds 47 °C for one minute during bone drilling. To solve this problem, this work studied a new thermal model by using three drilling parameters: drill diameter, feed rate and spindle speed. Effects of those parameters to heat generation were studied. The drill diameters were 4 mm, 6 mm and 6 mm; the feed rates were 80 mm/min, 100 mm/min and 120 mm/min whereas the spindle speeds were 400 rpm, 500 rpm and 600 rpm then an optimization was done by Taguchi method to which combination parameter can be used to prevent thermal necrosis during bone drilling. The results showed that all the combination of parameters produce confidence results which were below 47 °C and finite element analysis combined with Taguchi method can be used for predicting temperature generation and optimizing bone drilling parameters prior to clinical bone drilling. All of the combination parameters can be used for surgeon to achieve sustainable orthopaedic surgery.

1. Introduction

In orthopaedics surgery, bone drilling is widely known to create hole for insertion of screw [1]. Before the insertion of screw, drilling then tapping are conducted [2]. About 95% of post-trauma therapies need drilling holes that are needed to install the screws for plate attachments or other prosthetics or for fixing and correcting fractures of bone [3]. Study on bone drilling has retrieved many attentions. One significant aspect of the study concerns the generation of heat during the drilling process [4]. A few surgical methods have been explained in terms of influences during osteotomy of implant site [5].

The heat generation in bone drilling could be partially dissipated by the fluids of tissue or blood and partially involved by the chips created [6]. Cutting action between drill bit and bone results in significant heat because of removal of material and friction coefficient between the drill bit and bone [7]. Drilling actions when preparation of site of dental implant could cause not only physical failure to the bone but also a rise of temperature in surrounding the implant site [8]. Temperatures over 47 °C leads to thermal osteonecrosis providing to loosening of screw and subsequently refractures and failures of implant [9], reduces protein regeneration capacity [10] and irreversible bone death [11]. Bone drilling factors that affect the temperature change when the preparation of implant site including drill bit geometry, applied force, spindle speed [12], drill wear, the effect of saline temperature applied for irrigation and guide of drill were studied in several investigations resulting substantial importance [13].

Temperature measurement and distribution at the interface of drill bit and bone is hard. Engineers and scientists have been depeloving numerical method for heat transfer analysis. Finite Element

Analysis (FEA) shows effective proficiency in all engineering areas including in medical science. Bone drilling experiment is expensive, complex and time consuming. To cope this issue, finite element simulation provides benefits for analytical results, prediction and verification.

In 1999, Davidson performed a parametric analysis to study the thermal impact on bone drilling. Parameter used were drill diameter, spindle speed, feed rate, drill helix and point angle, density, bone thermal conductivity and specific heat. Highest temperature increased by feed rate from 0.45 to 1.8 mm/s then decreased slightly by feed rate of 4.5 mm/s and lowest temperature occured at helix angle of 38° and point angle of 130° [14]. In 2004, Li et al. compared surface heating and volumetric sources in the laser melting modeling for ceramic materials concluding that the simulation result with volumetric heating source was accurate and stable [15]. In 2011, Lee et al. worked out a new thermal model for bone drilling by a sensitivity analysis and single parametric study [16] followed by Mokhtar and Fawad in 2012 which was to accomplish theory of bone drilling modeling to gain the heat flux [17]. In 2015, Gok et al. studied the drilling and thrust power including coefficients of heat transfer for bone drilling numerical simulations using a K-wire. DEFORM-3D software was used to conduct the bone drilling process resulting there was a great consistency between FEA and experimental results. This has shown the finite element capability [18].

Other finite element method functions are to investigate drilling mechanism and to predict temperature of drilling. In 2009, Tu et al. applied a three-dimensional finite element model showing that
temperature decrease was caused by decreasing initial temperature of cutting tool and the thermal influenced zone size [19]. In 2013, Tu et al used an elastic–plastic finite element model for analyzing the behaviour of thermal contact between a drill bit and bone. The results showed the distribution of and rise of temperature rise surrounding the hole drilled could be successfully assessed [20]. In 2012, Sezek et al. analysed change of temperature during drilling of bone by application of Finite Elements Method (FEM) by deciding the points drilling in bone and found a secure zone or below temperature of 45 °C [21]. The numerical method in bone drilling simulation could be a foundation for surgical tools optimization including for determination of their properties [22] and provide tools for development of surgical operations [2].

Although many numerical simulations have been conducted by proposing new models, optimization of finite element analysis in bone drilling is still lacking. This paper presents an elaborated study from fundamental to optimization of three-dimensional model. A method is proposed to develop a thermal model to predict rise of temperature during bone drilling. In this study, effects of drill diameter, feed rate and spindle speed on generation of temperature in the bone drilling simulation were investigated then optimized using the Taguchi experimental design method. To optimize the effects of the process parameters, signal-to-noise (S/N) ratio and orthogonal array were calculated.

2. Heat generation

Coupled temperature-displacement method is applied to calculate and solve concurrently temperature and stress/displacement problems. A coupled analysis is applied as thermal and mechanical solutions influence strongly one another. Abaqus/Explicit software allows coupled temperature-displacement analysis modes. Fully coupled thermal-stress analysis in the software is always transient. In the software, equations of heat transfer are calculated by integration of explicit forward-difference time below [23].

$$\theta^N_{(i+1)} = \theta^N_{(i)} + \Delta t_{(i+1)} \dot{\theta}^N_{(i)} \tag{1}$$

Where θ^N as the temperature generated at node N whereas the subscript i as the number of increment in a step of explicit dynamic. The current temperatures are referred by values of $\dot{\theta}^N_i$ from the foregoing increment. The values of $\dot{\theta}^N_i$ are calculated at the start of the increment as follow [23].

(2)

$$\dot{\theta}^{N}_{(i)} = (C^{NJ})^{-1} + \left(P_i^J - F_i^J\right)$$

Where (C^{NJ}) as the lumped capacitance matrix, P^J as the vector of applied nodal source and F^J as the vector of internal flux. The explicit method integrates by means of time by applying a lot of small time increments. The forward-difference and central-difference operators are stable. The limit of stability for the two operators is obtained by [23]:

$$\Delta t \leq min\left(\frac{2}{\omega_{max}}, \frac{2}{\lambda_{max}}\right) \tag{3}$$

Where ω_{max} as the maximum frequency in the equations system of the response of mechanical solution and λ_{max} as the largest eigenvalue in the equations system of the response of thermal solution. An approximation to the limit of stability for the forward-difference operator in the response of thermal solution is calculated by [23]:

$$\Delta t \approx \frac{L_{min}^2}{2a} \tag{4}$$

Where L_{min}, as the minimum dimension of element in the mesh and $a = \frac{k}{\rho c}$ as the thermal diffusivity of the material. The letters k, ρ and c are the thermal conductivity, density, and specific heat of material, serially.

Boundary conditions may be applied to determine displacements/rotations and temperatures at nodes in fully coupled thermal-stress analysis. Boundary conditions at a step of dynamic coupled temperature-displacement response should apply suitable references of amplitude if boundary conditions are determined for the step without references of amplitude, they are applied instantly at the step start. Due to the software does not recognize jumps in displacement, a boundary condition of non-zero displacement value that is determined without reference of amplitude will be neglected and a boundary condition of zero velocity will be implemented [23]. The heat generated in the primary zone of shear deformation is calculated by [24]:

$$\dot{Q}_{sh}(r) = F_s(r)V_s(r) = \tau_s(r)A_s(r)\,V_s(r) \tag{5}$$

Where F_s is the force of shear along the plane of shear, V_s is the velocity of shear, τ_s is the yield stress of shear of bone, and As is the area of shear. They all are functions of drill bit radius r.

3. Method

3.1. Thermal model and simulation

A three-dimensional dynamic coupled temperature-displacement finite element analysis was used to simulate cortical bone drilling process. Simulations were carried out using Abaqus 6.14 software (Dassault Systemes Simulia Corp, USA). The bone drilling process was modeled in Abaqus/Explicit mode. Due to the simulation was dynamic, damage failure criteria was applied to manage the removal of element. Drill bit and cortical bone were modelled. The cortical bone was modeled as a rectangular work piece by a length of 30 mm, a width of 30 mm and a thickness of 6 mm. The drill bit was modeled with a helix angle of 23° and a point angle of 120°. Since drill bit stiffness is much higher than that of the bone and in order to reduce the analysis time, rigid body was assigned to the drill bit. For appropriate prediction of temperature during drilling, fine mesh was used at and in the surrounding of the cortical bone to be drilled, whereas coarse mesh was used to discretize the cortical bone away from the drill bit path. The cortical bone was meshed with C3D8T 8-node 110070 elements consisting of 61830 hexahedral elements and 48240 wedge elements whereas the drill bit was meshed with 6227 C3D4 4-node tetrahedron elements. Due to cortical bone has relatively small mechanical anisotropy, isotropic equivalent homogeneous material was applied instead of anisotropic and heterogeneous structure [25]. The interest region was the surrounding of the drilled hole where the temperature was highest. The reference point was located at the tip of the drill bit as the origin of the coordinate system and In order to model the rotations and feeds of the drill bits as shown in Figure 1.

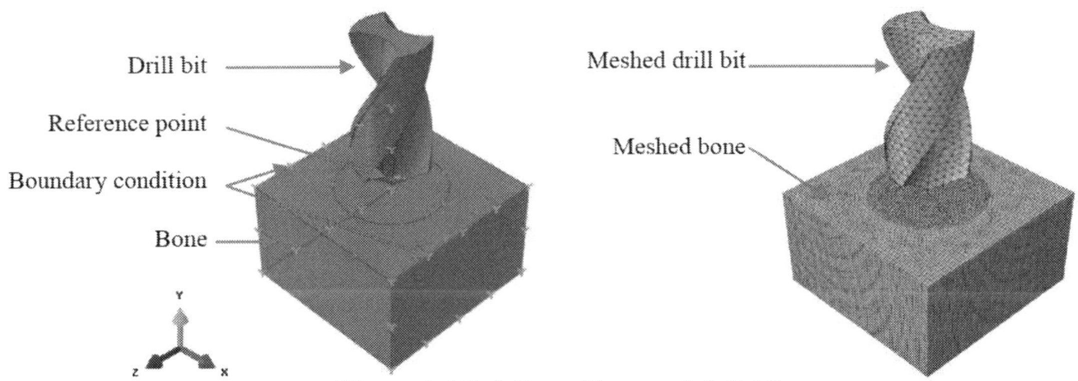

Figure 1. Modeling of bone and drill bit.

To guarantee stability and convergence of the simulation, a time period step of 0.0005 s was used. Thermal properties of the drill bit and the bone are listed in Table 1. Boundary condition was fixed on both vertical faces of the bone and for the drill bit was set free for rotation along the bone axis and vertical movement. Initial temperatures for the bone and the drill bit were set at 23 °C following room temperature. These initial and boundary temperatures were the same for all simulations. Contact interaction type of the drill bit and the bone was surface-to-surface contact with friction coefficient of 0.7. First surface (master) was assigned to the drill bit, whereas second surface (slave) was assigned to the cortical bone.

Table 1. Properties of drill bit and bone

Parameter	Drill bit (HSS)	Bone
Thermal conductivity, W/m °C	24 [26]	0.5 [27]
Density, kg/m^3	8150 [26]	2000 [27]
Young modulus, GPa	230 [26]	34.3 [28]
Poisson's ratio	0.3 [29]	0.36 [18]
Expansion, /°C	11.6 x 10^{-6} [26]	2.7 x 10^{-5} [30]
Specific heat, J/kg °C	420 [26]	1290 [27]

Johnson-cook damage model was used for the chip - workpiece removal in the drilling simulations. [31].

$$\bar{\sigma} = (A + B(\varepsilon^{-pl})^n)\left[1 + Cln\,\frac{\dot{\varepsilon}^{-pl}}{\dot{\varepsilon}_0}\right](1 - \hat{\theta}^m) \tag{6}$$

Where $\bar{\sigma}$ is the equivalent flow stress, A is the material initial yield strength, the parameters B, n, C (as listed in table 2) and m are material model parameters, ε^{-pl} is the equivalent plastic strain and $\dot{\varepsilon}^{-pl}$ is the rate of equivalent plastic strain which is normalized with a rate of reference strain $\dot{\varepsilon}_0$ and $\hat{\theta}$ is the homologus temperature calculated in equation 7 [31].

$$\hat{\theta} = \frac{(\theta - \theta_{room})}{(\theta_{melt} - \theta_{room})} \tag{7}$$

Where θ is the workpiece instantaneous temperature, θ_{room} is the room temperature and θ_{melt} is the material melting temperature and θ_{melt} is the material melting temperature.

Table 2. Johnson-Cook properties of drill bit and bone

	A (MPa)	B (MPa)	*C*	*n*
Drill bit [29]	353.4	102.6	0.21	0.29
Bone [32]	50	101	0.03	0.0

Element failure has disadvantage which is material deletion. The main disadvantage with element failure is the deletion of material. It is mostly non-physical that mass is deleted from the drilling process and this deletion affects the forces (pressure) between the drill bit and the bone. Therefore, it affects all simulation result. To minimize this effect, the density of mesh must be very fine [31].

3.2. Taguchi method

The Taguchi method is the most significant technique for effective and systematic approach for better optimization solution proposed by Taguchi (1990) [33] and to provide a significant reduction in experiment number for same number of process control factors [34]. Smaller-the-better criterion was used to calculate signal-to-noise (S/N) ratio where T is temperature generated, then response was calculated using the obtained S/N ratio.

$$S/N\ Ratio = -10 \log T^2 \tag{8}$$

In Taguchi method, for conducting the experiment, factors and levels were defined. Factors are controllable parameters which affect the experiment significantly. Three principal drilling parameters were chosen for numerical calculations same as the situation of real drilling processes. They were drill bit diameter, feed rate and spindle speed,. The bone temperature was the target during the drilling process, so the maximum temperature of the entire bone was extracted. The drill diameter, feed rate and spindle speed were the controllable parameters for this experiment since they can influence temperature generated during bone drilling. The controllable parameters along with the levels used are shown in Table 3.

Table 3. Drilling parameter

Parameter	Level I	Level II	Level III
Drill bit diameter (mm), X	4	6	8
Feed rate (mm/min), Y	80	100	120
Spindle speed (rpm), Z	400	500	600

Table 4. Design matrix for L_9 orthogonal array

Experiment	X	Y	Z
1	4	80	400
2	4	100	500
3	4	120	600
4	6	80	500
5	6	100	600
6	6	120	400
7	8	80	600
8	8	100	400

9	8	120	500

After defining the factors and their levels, orthogonal array was defined. To select an appropriate orthogonal array, a degree of freedom (DOF) was calculated. DOF is needed to determine which level is better and is defined as the number of comparisons. DOF was 1 for mean value and 8 for two each for the remaining factors, so the total DOF was 9. Therefore L_9 orthogonal array or a number of nine experiments were selected for this study as shown in Table 4.

4. Results and discussion

Figure 2 shows heat generation during drilling simulation. Maximum temperature resulted from each experiment was used for the optimization. For calculating S/N ratio, the objective function of the temperature generated was smaller-the-better criterion due to the lower the temperature, the better the bone drilling process. The S/N ratios of all the experiments were calculated and tabulated as shown in Table 5. The average of all the factors at each level or response was then calculated from the obtained S/N ratio as shown in Table 6.

Figure 2. Heat generation during drilling.

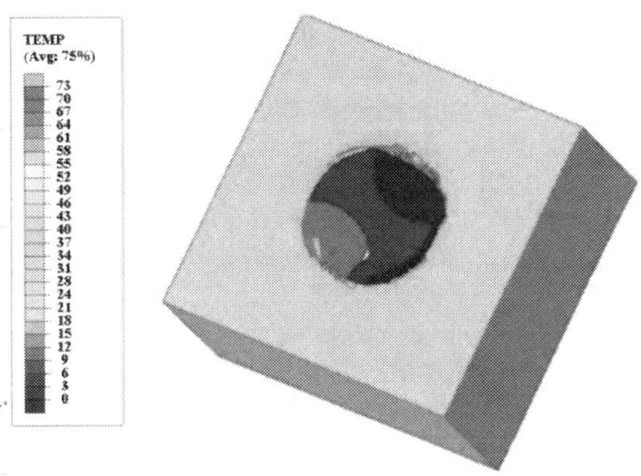

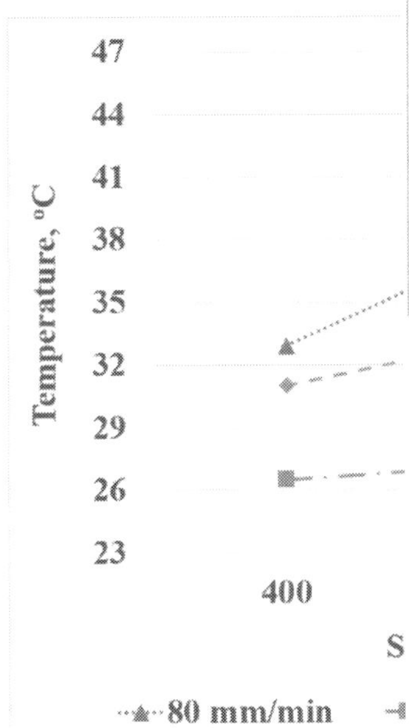

Figure 3. Graph of temperatures generated at each factor and level.

Table 5. Temperature and S/N ratio

Experiment	X	Y	Z	Temperature °C	S/N Ratio
1	4	80	400	32.9	-30.34
2	4	100	500	38.6	-31.73
3	4	120	600	43.5	-32.77
4	6	80	500	26.5	-28.46
5	6	100	600	27.3	-28.72
6	6	120	400	31.6	-29.99

7	8	80	600	31.0	-29.83
8	8	100	400	33.6	-30.53
9	8	120	500	35.6	-31.03

Table 6. Response table for S/N ratio

Parameter	Level 1	Level 2	Level 3	Delta	Rank
X	-31.62	-29.06	-30.46	1.15	2
Y	-29.55	-30.33	-31.26	1.72	1
Z	-30.03	-30.52	-30.59	0.56	3

This section discusses the result obtained by using methodology explained in the previous section. The variation in temperature generated are shown in Figure 3. Temperature increases as the spindle speed increase regardless of the increasing of the feed rate. The feed rate of 80 mm/min generates highest temperature followed by the feed rate of 120 mm/min and 100 mm/min. The change effect in the spindle speed indicates a significant increase in the temperature generated due to with the increasing of spindle speed, bone chips may erode the drilled surface with more centrifugal force. The change effect in the feed rate indicates a decrease in the temperature generated due to by increasing the feed rate, contact time between the drill bit and the bone is greatly reduced [35]. Temperature decreases when the feed rate changes from 80 mm/min to 100 mm/min then increases when the feed rate changes from 100 mm/min to 120 mm/min. However, the results show that all temperature generated are below 47 °C.

According to the calculation results from the Taguchi method that are shown in Table 6, the combination of drilling parameters that produces the best temperature generated is by the drill diameter of 6 mm, the feed rate of 80 mm/min and the spindle speed of 400 rpm. However all combination can be used for drilling process since those parameters produce temperatures which are below 47 °C.

5. Conclusions

The present paper used finite element analysis for temperature generation then the results were optimized by Taguchi method. The above results and discussions are summarized as follows:

- From the three linear graphs, it is clear that the optimum values of the factors and their levels are the diameter of 6 mm, the feed rate of 80 mm/min and the spindle speed of 400 rpm.
- The results show that the drill diameter, the feed rate and spindle speed give significant effects on temperature generation in bone drilling. It is found that feed rate has the highest influence on temperature generated followed by drill diameter then the spindle speed respectively.
- All the factors and levels used can be reference for parameters used by surgeon in bone drilling to prevent thermal necrosis.
- Combination of finite element analysis with Taguchi method is suitable for analysis prior to clinical bone drilling since experimental bone drilling needs complex tools and preparation.

Acknowledgments

Authors thank to Mechanical Engineering Department of Universiti Teknologi PETRONAS and Malaysia of Higher Education for supporting this project by providing Fundamental Research Grant Scheme (FGRS), no. FRGS/2/2014/TK01/UTP/02/2).

Conflict of interest

All authors agree that there is no any conflict of interest for this manuscript.

References

[1] Pandey R K and Panda S S 2015 Optimization of multiple quality characteristics in bone drilling using grey relational analysis *Journal of Orthopaedic* **12(1)** 39-45.

[2] Lughmani W A, Marouf K B and Ashcroft I 2015Drilling in cortical bone a finite element model and experimental investigations *Journal of The Mechanical Behavior of Biomedical Materials* **42** 32–42.

[3] Accini F, Díaz I and Gil J J 2016 Using an admittance algorithm for bone drilling procedures," *Computer Methods and Programs in Biomedicine* **123** 150-158.

[4] Feldmann A, Wandel J and Zysset P 2016 Reducing temperature elevation of robotic bone drilling *Medical Engineering and Physics* **38(12)** 1495-1504

[5] Mohlhenrich S C, Abouridouane M, Heussen N, Holzle F, Klocke F and Modabber A 2016 Thermal evaluation by infrared measurement of implant site preparation between single and gradual drilling in artificial bone blocks of different densities *Int. J. Oral Maxillofac. Surg.* **45(11)** 1478-1484.

[6] Li X, Zhu W, Wang J and Deng Y 2016 Optimization of bone drilling process based on finite element analysis *Applied Thermal Engineering* **108(5)** 211-220.

[7] Gupta V and Pandey P M 2016 Experimental investigation and statistical modeling of temperature rise in rotary ultrasonic bone drilling *Medical Engineering and Physics* **38(11)** 1330-1338.

[8] Kim S J, Yoo J, Kim Y S and Shin S W 2010 Temperature change in pig rib bone during implant site preparation by low-speed drilling *J Appl Oral Sci.***18(5)** 522-527.

[9] Augustin G, Davila S, Mihoci K, Udiljak T, Vedrina D S and Antabak A 2008 Thermal osteonecrosis and bone drilling *Archives of Orthopaedic and Trauma Surgery* **128(1)** 71-77.

[10] Krause W 1987 Orthogonal bone cutting: saw design and operating characteristics *J. Biomech. Eng.* **109(3)** 263-271.

[11] Lundskog J 1972 Heat and bone tissue: an experimental investigation of the thermal properties of bone and threshold levels for thermal injury *Scand J. Plast Reconstr Surg.* **9** 1-80.

[12] Soriano J, Garay A, Aristimuño P and Arrazola P 2014 Study and improvement of surgical drill bit geometry for implant site preparation *The International Journal of Advanced Manufacturing Technology* **74(5)** 615-627.

[13] Misic T, Markovic A, Todorovic A, Colic S, Miodrag S and Milicic B 2011 An in vitro study of temperature changes in type 4 bone during implant placement: bone condensing versus bone drilling *Oral Surg. Oral Med. Oral Pathol. Oral Radiol. Endod.* **112** 28-33.

[14] Davidson S R 1999 Heat transfer in bone during drilling.

[15] Li J, Li L and Stott F 2004 Comparison of volumetric and surface heating sources in the modeling of laser melting of ceramic materials *International Journal of Heat and Mass Transfer* **47(6-7)** 1159-1174.

[16] Lee J, Rabin Y and Ozdoganlar O B 2011 A new thermal model for bone drilling with applications to orthopaedic surgery *Medical Engineering & Physics* **33(10)** 1234-1244.

[17] Mokhtar M and Fawad H 2012 Bone drilling modeling and simulation techniques *Proceedings of IEEE Symposium on Business, Engineering and Industrial Applications (ISBEIA), Bandung, Indonesia.*

[18] Gok A, Gok K and Bilgin M 2015 Three-dimensional finite element model of the drilling process used for fixation of Salter-Harris type-3 fractures by using a K-wire *Mechanical Sciences* **6** 147-154.

[19] Tu Y, Liu Y, Yang W, Chen L, Hong Y, Chen Y and Lin L 2009 Temperature rise simulation during a kirschner pin drilling in bone *Proceedings of the 3rd IEEE International Conference on*

Bioinformatics and Biomedical Engineering, Beijing, China.

[20] Tu Y K, Chen L W, Ciou J S, Hsiao C K and Chen Y C 2013 Finite element simulation of bone temperature rise during bone drilling based on a bone analog *Journal of Medical and Biological Engineering* **33(3)** 269-274.

[21] Sezek S, Aksakal B and Karaca F 2012 Influence of drill parameters on bone temperature and necrosis: a FEM modelling and in vitro experiments *Computational Materials Science* **60** 13-18.

[22] Paszenda Z and Basiaga M 2009 FEM analysis of drills used in bone surgery *Archives of Materials Science and Engineering* **36(2)** 103-109.

[23] 2014 Abaqus Manual User Guide, version **6.14.** Dassault Systèmes Simulia Corp.

[24] Sui J, Sugita N and Mitsuishi M 2015 Thermal Modeling of Temperature Rise for Bone Drilling with experimental validation *Journal of Manufacturing Science and Engineering* **137(6)** 061008.

[25] Li S, Demirci E and Silberschmidt V V 2013 Variability and anisotropy of mechanical behavior of cortical bone in tension and compression *J. Mech. Behav. Biomed. Mater.* **21** 109-120.

[26] Taylor High Speed Steel Ltd.

[27] Tai B L, Palmisano A C, Belmont B, Irwin T A, Holmes J and Shih AJ 2015 Numerical evaluation of sequential bone drilling strategies based on thermal damage *Medical Engineering and Physics,* **37(9)** 855–861.

[28] Khan S, Warkhedkar R and Shyam A 2014 Human Bone strength Evaluation through different Mechanical Tests *International Journal of Current Engineering and Technology.*

[29] Gök K, Türkes E, Neseli S, Saglam H and Gök A 2013 The validation as experimental and numerical of the values of thrust force and torque in drilling process *Journal of Engineering Science and Technology Review* **6(3)** 93-99.

[30] Ranu H S 1987 The thermal properties of human cortical bone: an in vitro study *Engineering in medicine* **16(3)**.

[31] Zouhar J and Piska M 2008 Modelling the orthogonal machining process *MM science Journal* 49-52,

[32] Alam K, Mitrofanov A V and Silberschmidth V V 2010 Thermal analysis of orthogonal cutting of cortical bone using finite element simulations *Journal of Experimental and Computational Biomechanics (IJECB)* **1(3)** 236-251.

[33] Basavarajappa S, Chandramohan G and Davim J P 2007 Application of Taguchi techniques to study dry sliding wear behaviour of metal matrix composites *Mater. Des.* **28** 1393-1398.

[34] Roy R 2001 Design of experiments using the Taguchi approach: 16 steps to product and process improvement ISBN: 978-0-471-047136101-5.

[35] Singh G, Jain V, Gupta D and Gha A 2016 Optimization of process parameters for drilled hole quality characteristics during cortical bone drilling using Taguchi method *Journal of The Mechanical Behavior of Biomedical Materials* **62** 355-365.

Investigation of Carbon Nanotube Ink with PDMS Printing Plate on Fine Solid Lines Printed by Micro-flexographic Printing Method

S Hassan[1a], M S Yusof[2a], S Ding[3b], M I Maksud[4a], M N Nodin[5a], K A Mamat[6a], M S Sazali[7a], M Z Rahim[8a]

Faculty of Mechanical and Manufacturing Engineering,
Universiti Tun Hussein Onn Malaysia,
86400 Batu Pahat, Johor, Malaysia[a]

School of Aerospace, Mechanical and Manufacturing Engineering,
RMIT University East Campus,
Melbourne, VIC 3083, Australia[b]

Email: suhaimihas@uthm.edu.my[1], mdsalleh@uthm.edu.my[2], songlin.ding@rmit.edu.au[3], midris1973@gmail.com[4], mdnor.nodin@yahoo.com[5], khairul8906@gmail.com[6], iwan16289@yahoo.com[7], zulafif@uthm.edu.my[8]

Abstract. Micro-flexographic printing which is combination of flexography and micro-contact printing is a unique printing technique that create graphic, electronic and biomedical printed image on variable substrates. Therefore, an investigation for its potential usage in global area of printing hence a research for suitable ink and printing plate related to this process is vital. This paper will focus on investigating the printability of Carbon Nanotube (CNT) ink with polydimethylsiloxane (PDMS) printing plate. There are two types of CNTs ink which are solvent base and water base. A printing trial has been carried out in comparison with CNTs inks and PDMS printing plate with fine solid lines image on it to check the suitability of the printing plate development for micro-flexographic printing technique. The result shows that CNT water base ink is suitable with PDMS printing plate in printing fine solid lines image.

1. Introduction

Recently, researchers have shown that printing process play an important roles in producing the printed fine solid lines image expecially in electronic industry. The research done by Zhang has investigated the manufacturability of organic electronics like organic field effect transistors (OFETs) using commercially available printing technologies and materials systems qualified for use in microelectronic products. The organic integrated circuits (IC) may be the low cost solution for driving electronic devices like smart cards, RFID tags, flexible displays and personal area networks. The material system selection for transistor structures and active layers is based on printing technology requirements [1].

IOP Conf. Series: Materials Science and Engineering **203** (2017) 012017 doi:10.1088/1757-899X/203/1/012017

Flexography printing process is a direct rotary printing that uses moulded printing die known as printing plate which is made from rubber or photopolymer. In a normal flexography machine design, there are multiple wheels that work in unison to pick up ink from an ink reservoir as shown in figure 1. Then, the ink will be delivered to printing die or printing plate. Lastly, printing plate will roll the ink onto product being mark or substrate [2].

Other than flexography printing method, micro-contact printing (μCP) is a famous printing technique in producing product in micro or nano scale pattern. Micro-contact printing (μCP) is performing by release patterns on a master polydimethylsiloxane (PDMS) stamp to form patterns of self-assembled monolayers (SAMs) of ink on the surface of a substrate through conformal contact like showing in figure 2. It is a low cost technique for the preparation of micro structure printing surface. PDMS is a suitable material for μCP stamp because it has number of properties which well suit for μCP process [3].

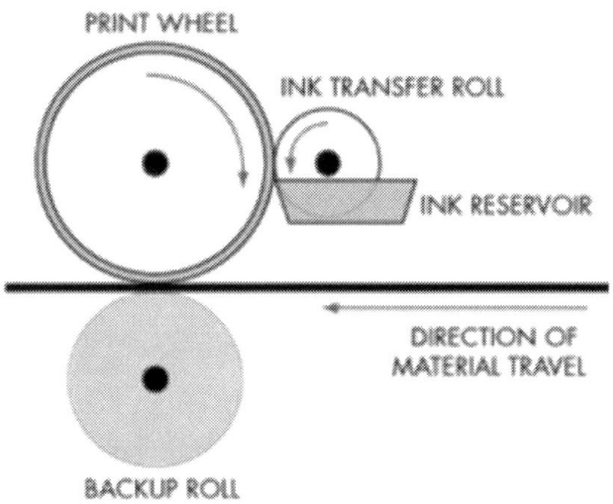

Figure 1: Flexography printing technique

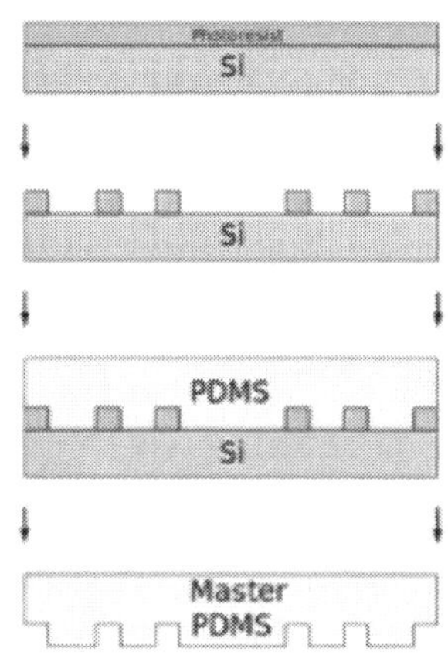

Figure 2: Micro-contact printing printing technique

Low cost material and production in producing electronic parts by printing process like flexography is very important. The photopolymer material as a printing plate in flexography printing process is unable to produce micron scale lines of features. Hence, PDMS can be used to replace the photopolymer as a new printing plate in flexography printing process. Printing trial has been conducted by previous researcher in using PDMS printing plate for printing fine line in micro scale pattern [4].

The printability of CNTs as a conductive ink had been performed by previous researcher in flexography printing process. The author managed to print fine solid line image pattern by the CNT ink on variable substrates such as silica, biaxially oriented polypropylene (BOPP), textile and paper. Simple experimental test of ink functional performance had been done by lighting up light emitting diode (LED) lamp. The author concluded that CNTs water base was the best ink for printing on variable substrates but maintains high electric conductivity [5].

The flexography printing method had a capability in printing fine solid lines image. The previous research [6] used roll to roll flexography printing technique which was web press industrial method, the author managed to print out fine solid lines image by using carbon graphic inks. This technique used photopolymer as a printing plate to transfer the ink from plate roller to substrate. It was a step forward of developing high speed printing in electronic with simple, rapid, low cost method, less waste and roll-to-roll capability [7].

The application of flexography printing process in producing fine solid lines image in micro or nano scale is not limited to create electronic part only. This printing technique can be also applied in

IOP Conf. Series: Materials Science and Engineering **203** (2017) 012017 doi:10.1088/1757-899X/203/1/012017

biomedical industry [8]. By utilizing the PDMS printing plate, cell culture can be printed onto flexible polymer substrate for bio-sensor making [9].

2. Research Methodology

In this study, there were two type of CNTs ink used for the fine solid lines printing trial. First ink was CNTs solvent base liked showing in figure 3. CNTs solvent base ink had 90% to 95% chemical composition of polypropylene glycol [5]. Second ink was CNTs water base as shown in figure 4. This type of ink had 80% to 90% composition of water. Both of the inks were supplied by Fuji Ink Industries Limited.

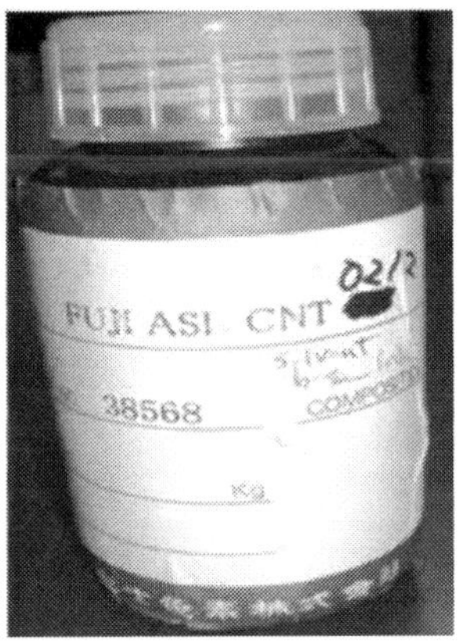

Figure 3: Carbon Nanotube (CNT) solvent base ink

Figure 4: Carbon Nanotube (CNT) water base ink

A pattern of multiple fine solid lines image of PDMS printing plate was prepared as shown in figure 5. PDMS printing plate had an advantage in producing micro, even nano-scale size in fine solid lines printing [10]. The PDMS was poured into the mould with the silicon wafer as a master image plate to get fine solid lines image with different size of width and distance gap. Figure 6 showed the PDMS printing plate with multiple fine solid lines image. The PDMS printing plate size was approximately 30 mm x 20 mm for length and width and 1mm thickness.

Figure 5: Silicon wafer master image

Figure 6: PDMS printing plate

The printing process was started with the preparation of customized micro-flexographic printing machine in labaratory scale as shown in figure 7. Micro-flexographic printing machine was a combination of flexography and micro-contact printing method [11]. Flexography was one of the fastest printing processes but it had limitation in micro to nano printing scale. Compared with micro-contact printing, it was very slow but can achieved micro to nano fine solid lines image. This combination printing method will have a high printing capability with high speed production, micro to nano scale image pattern and low cost production process.

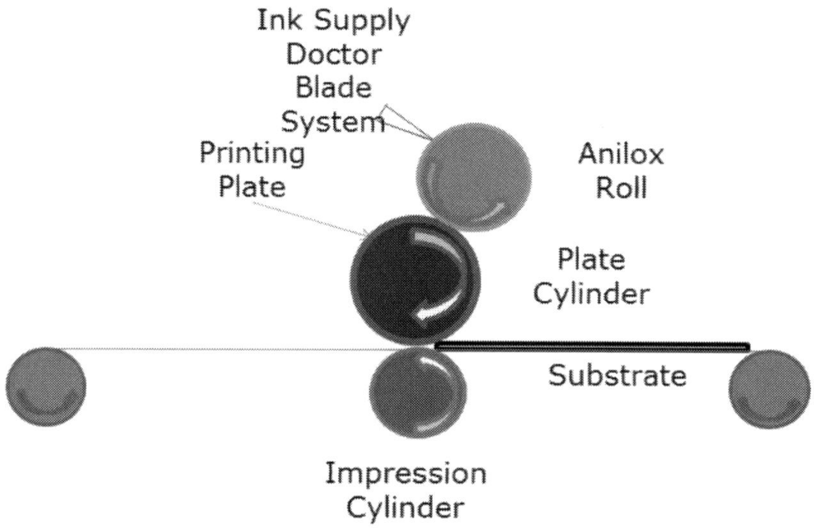

Figure 7: Schematic diagram of the micro-flexographic printing process

In this customize design apparatus, basic components of flexography printing machine liked impression cylinder, plate cylinder, anilox roll and doctor blade were replicated. Micro-flexographic printing process used the same concept liked flexography printing method. PDMS printing plate with multiple fine solid lines image was attached to the plate cylinder. The CNT ink was transfer to the PDMS printing plate by using an engraved cylinder roll which known as anilox roll. The anilox roll was the main control of ink tranferred to the printing plate and substrate [12].

The printing capability was checked by adjusting printing parameters liked speed and roller engagement. All the parameters needed to be taken care during printing trial due to the aim of this project was very critical. In parameter setting approach, the speed was set at 15 rpm. All the parameters mention was set to get a good result in printing trial. The printed fine solid lines image was checked by common microscope known as Olympus BX60M. Technical specification of the microscope are 30mm/s speed, 0.1µm encoder resolution and below 2.0µm absolute accuracy. The result will show the direction of this study in meeting the objective.

2.1. The printed image analysis using microscope Olympus BX60M

The printed fine solid lines image was checked by using common microscope known as Olympus BX60M. Technical specification of the microscope were 30mm/s speed, 0.1µm encoder resolution and below 2.0µm absolute accuracy. This equipment was capable to check the line features including line width and gap.

3. Result and Analysis

Experimental process for printing trials had been done to find the suitability between CNTs ink and PDMS printing plate which could be selected as preliminary result. All the experiment was ran by using micro-flexographic printing process. First printing trial was done on PDMS printing plate by

IOP Conf. Series: Materials Science and Engineering **203** (2017) 012017 doi:10.1088/1757-899X/203/1/012017

using CNTs solvent base ink. From the result, it showed that PDMS Printing plate was not suitable to be used with CNT solvent base ink. Figure 8 showed the evidence which the PDMS was damaged when both of the printing part or parameter were used together. CNT solvent base ink was caused the damaged of the PDMS surface of fine solid lines image.

Figure 8: PDMS printing plate condition after being used with CNT solvent base ink

Second experiment of printing trial was done on PDMS printing plate by using CNT water base ink. The printing trial had been done successfully on paper substrate. The multiple fine solid lines pattern image printed result liked showing in Figure 9(a-d) depended on the viscosity setting. The ink viscosity was setting from low to high which was ratio between water and CNT ink. Figure 9(a) showed the lowest viscosity setting with unclear fine solid lines image. When the viscosity was increased, fine solid line pattern image will be more cleared as shown in Figure 9(d).

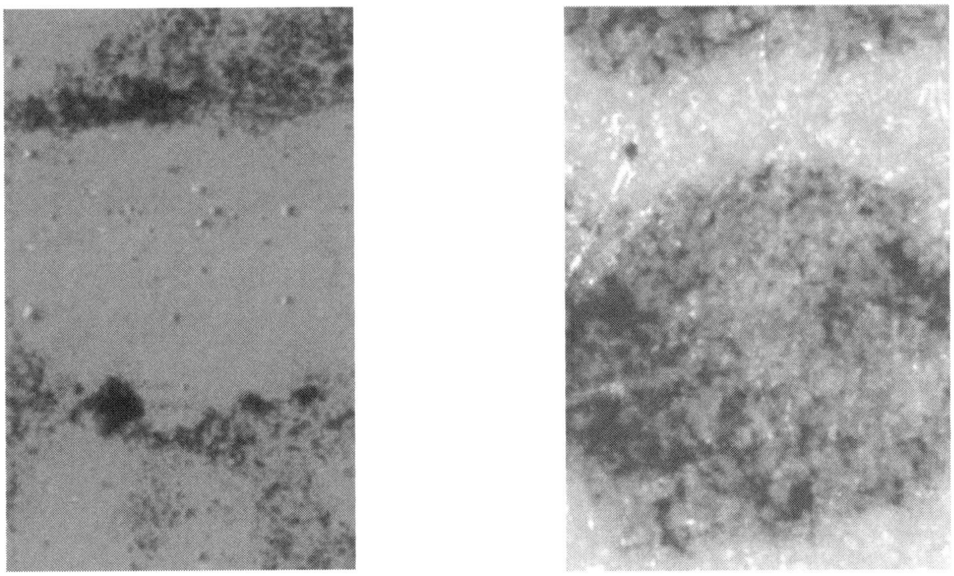

(a) (b)

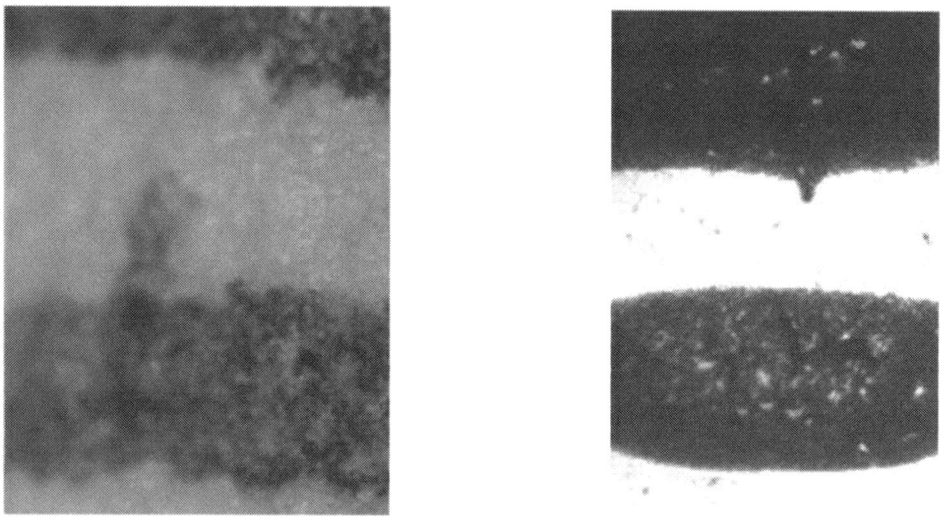

(c) (d)

Figure 9: Result of viscosity changing in using CNT water base ink on the paper substrate: (a) Low viscosity (b) Moderate-1 viscosity (c) Moderate-2 viscosity (d) High viscosity

This result showed that micro-flexographic printing technique needed a high viscosity setting to get a good image pattern compared with micro-contact printing which needed a low viscosity setting for printing ink [13]. Other parameters which had significant effect on micro-flexographic printing process were machine parameters setting and substrate. The examples of machine parameters setting were machine speed and engagement between rollers. Those parameters were well manipulated so that good result could be achieved.

The fine solid lines image from Figure 9(d) was taken as the best result for line measurement analysis. The final result from microscope Olympus BX60M measurement was shown that the fine solid lines printed image width and gap size was 0.35 mm and 0.25 mm respectively. The measurement result of the printed image was depended on the fine lines image size on PDMS printing plate surface.

During the experimental printing process, there were several major concern involved. The concerns were printing speed, engagement, type of ink, ink viscosity and printing plate material. Some ink liked CNT solvent base cannot be used with polymer printing plate and ink viscosity will afffect the best image result for measuring process analysis. Therefore, all those concerns needed to be investigated further in order to get better result in the future for printing development process.

4. Conclusion

From this study, several conclusions can be summaries as follow:

1. CNT solvent base ink is not suitable to use with PDMS printing plate.
2. CNT water base ink is suitable to be used with PDMS printing plate for printing purpose.
3. Ink viscosity is effected the image result. When the ink viscosity is increased, fine solid line printed image is more cleared and give good result for measuring process.
4. Micro-flexographic is a combination of flexographic and micro-contact printing has a good potential in printing fine solid lines image pattern for graphic, electronic and bio-medical purpose.
5. Micro-flexographic is a good candidate for fine solid lines printing with ink property, substrates and process parameters are main role to achieve the implementation.

Acknowledgments
This project was supported by 'Skim Latihan Akademik Bumiputera' (SLAB) from Universiti Tun Hussein Onn Malaysia, Ministry of High Education Malaysia.

References

[1] Zhang J, Brazis P, Chowdhuri A R, Szczech J and Gamota D 2002 Investigation of using contact and non-contact printing technologies for organic transistor fabrication, *MRS Proceedings*. Cambridge Univ Press pp P6. 3.
[2] Kipphan H, 2001. Handbook of print media: technologies and production methods. Springer.
[3] Kaufmann T and Ravoo B J 2010 Stamps, inks and substrates: polymers in microcontact printing *Polymer Chemistry* 1(4): 371-387.
[4] Maksud M, Yusof M and Jamil M 2014 An Investigation onto Polydimethylsiloxane (PDMS) Printing Plate of Multiple Functional Solid Line by Flexographic *Advanced Materials Research* 844: 158-161.
[5] Maksud M I, Yusof M S, Embong Z, Nodin M N and Rejab N A 2014 An Investigation on Printability of Carbon Nanotube (CNTs) Inks by Flexographic onto Various Substrates *International Journal of Materials Science and Engineering* Vol. 2(No. 1).
[6] Yusof M S 2011 Printing Fine Solid Lines in Flexographic Printing Process Swansea University Swansea 201 pp.
[7] Hassan S, Yusof M S, Maksud M I, Nodin M N and Rejab N A 2015 A Feasibility Study of Roll to Roll Printing on Graphene *Applied Mechanics and Materials* 799-800: 402-406.
[8] Hassan S, Yusof M S, Maksud M, Nodin M, Rejab N A and Mamat K 2015 A study of nano structure by roll to roll imprint lithography, *International Symposium on Technology Management and Emerging Technologies (ISTMET)*. IEEE pp 132-135.
[9] Maksud M, Yusof M and Mahadi Abd Jamil M 2013 Optimizing a polydimethylsiloxone (PDMS) into flexographic printing process for RFID biomedical devices and cell cultures, *Biomedical Engineering International Conference (BMEiCON), 2013 6th*. IEEE pp 1-4.
[10] Maksud M I, Nodin M N, Yusof M S and Hassan S 2016 Utilizing rapid prototyping 3D printer for fabricating flexographic PDMS printing plate *ARPN Journal of Engineering and Applied Sciences* 11(12): 7728-7734.
[11] Hassan S, Yusof M, Embong Z and Maksud M 2017 Surface adhesion study of La2O3 thin film on Si and glass substrate for micro-flexography printing, *AIP Conference Proceedings*. AIP Publishing pp 040006.
[12] Hassan S, Yusof M S, Embong Z and Maksud M I 2016 Angle resolved x-ray photoelectron spectroscopy (ARXPS) analysis of lanthanum oxide for micro-flexography printing *AIP Conference Proceedings* 1704: 040002.
[13] M.S.Yusof, M.I.Maksud and Z.Embong 2013 An Investigation Into Parameter Effect On Microcontact Printing Using Carbon And Biological Ink And XPS Study On Surface Morphology, *International Conference on Innovative Technologies, IN-TECH 2013* Budapest pp 329-332.

The Influence of Injection Molding Parameter on Properties of Thermally Conductive Plastic

N. Hafizah Azis[1], M. Zulafif Rahim[1,*], Nasuha Sa'ude[1], N. Rafai[1], M.S. Yusof[1,2], ALM Tobi[3], ZM Sharif[4], M. Rasidi Ibrahim[1], A.E Ismail[1]

[1]Faculty of Mechanical and Manufacturing Engineering, University Tun Hussein Onn Malaysia (UTHM) P.O. Box 101, Parit Raja, Batu Pahat 86400, Johor, Malaysia.

[2]Tensei Innovative Solution, No. 51 Block C, Universiti Tun Hussein Onn Malaysia (UTHM) City Campus, Parit Raja, Batu Pahat 86400, Johor, Malaysia.

[3]Faculty of Engineering Technology, University Tun Hussein Onn Malaysia (UTHM) P.O. Box 101, Parit Raja, Batu Pahat 86400, Johor, Malaysia.

[4]Universiti Tenaga Nasional, Seksyen 11, Selangor, 43650 Bandar Baru Bangi, Malaysia.

*zulafif@uthm.edu.my

Abstract. Thermally conductive plastic is the composite between metal-plastic material that is becoming popular because if it special characteristic. Injection moulding was regarded as the best process for mass manufacturing of the plastic composite due to its low production cost. The objective of this research is to find the best combination of the injection parameter setting and to find the most significant factor that effect the strength and thermal conductivity of the composite. Several parameter such as the volume percentage of copper powder, nozzle temperature and injection pressure of injection moulding machine were investigated. The analysis was done using Design Expert Software by implementing design of experiment method. From the analysis, the significant effects were determined and mathematical models of only significant effect were established. In order to ensure the validity of the model, confirmation run was done and percentage errors were calculated. It was found that the best combination parameter setting to maximize the value of tensile strength is volume percentage of copper powder of 3.00%, the nozzle temperature of 195°C and the injection pressure of 65%, and the best combination parameter settings to maximize the value of thermal conductivity is volume percentage of copper powder of 7.00%, the nozzle temperature of 195°C and the injection pressure of 65% as recommended.

1. Introduction

Plastic material currently used in industries such as automotive, manufacturing and aerospace. Increase demand for light weight and easy to shape of the component is main selected criteria to reduce energy consumption, lower manufacturing cost and time. Therefore, the best of material for suitable to be used to meet these criteria is plastic. Thermoplastic is a main types of plastic. This plastic can be reused after re-melting and the chemical structure remain unchanged during heating and shaping process. The example for thermoplastic is acrylic, ABS (acrylonitrile butadiene styrene), nylon, and PP (polypropylene). There are some of the noteworthy properties of thermoplastic is have high strength, are lightweight, good chemical and stress cracking resistant, excellent high temperature stability, heat resistance and have relatively low processing cost [1].

But the use of thermoplastic is limited to certain application due to its low thermal conductivity and heat resistant. Thermal conductivity of thermoplastic was reported around 0.2 W/mK (Watts/meter-°Kelvin), considerably lower than most thermally conductive plastic compounds that typically have 10 to 50 time higher conductivity (1-10 W/mK) [2]. Therefore, plastic modification is a way to improve their thermal conductivity and strength through the use of additive or filler [3, 4]. There are several kind of additive that usually used which include metal, ceramic and polymer [5]. Injection molding, extrusion, and compression molding are the methods that is normally implemented to produce thermoplastic composites with additive material [4].

2. Methodology

Copper powder, ABS and surfactant agent are the materials used in this conductive plastic fabrication. ABS considered as an environment friendly material as they are completely recyclable. The properties of material are; ABS material: density of 1.03 g/cm^3 and melting point of 230°C; copper powder: purity of 99%, particles size distribution of 17 μm – 50 μm, melting temperature of 1080 °C and the density of 9.84 g/cm^3. The percentage of volume distribution for the copper powder, ABS and paraffin wax is shown in Table 1

Table 1: Percentage of volume distribution for the copper powder, ABS and paraffin wax.

No	Composition volume, %			Density , g/cm^3
	Copper	ABS	Paraffin wax	
1	3	96	1	1.27
2	7	90	3	1.58

In this experiment, the actual weight of compounding was determined and calculated by the following equation:

$$Wc = Wcop + W\ ABS / (1 - Ws\ \%) \tag{1}$$

Where;
W_c = weight of composite material in grams
W_{copper} = weight of copper powder in grams
W_{ABS} = weight of ABS in grams
$W_S\%$ = weight percentage (wt %) of surfactant

Table 2 shows the calculated elemental and total weight of ABS-Cu composite.

Table 2: Total weight of ABS-Cu composite

No	Weight, g			Total weight, g
	Cooper, g	ABS, g	Paraffin wax, g	
1	21.19	78.11	0.71	100.00
2	39.66	58.2	1.71	100.00

ABS material was crushed and sieved into approximately 1mm-5mm particles size purposely to eliminate the impurities content. In order to achieve a homogeneous condition, the powder was mixed with the surfactant agent. The process was carried out by Brabender Plastograph mixer, type W50 at 185°C with the speed of 20 rpm for 30 minutes. Next, the prepared feedstock was crushed by granulator machine to produce pallets with similar size. The injection process of the pallets was done by horizontal NPF-1F molding machine to produce the sample for tensile strength and thermal conductivity analysis.

The injection machine specification are; screw diameter of 19 mm, injection capacity of 14 cm^3, injection rate 20% and injection pressure of 60% to 65%. In order to prevent the melt from sticking at the screw, the barrel temperature was started from low to high temperature. Figure 1 and Figure 2 show, the injection molding machine and the zone temperature in injection molding machine respectively. The machine barrel temperatures that consist of five temperature zone were set as follow; the nozzle temperature was in the range of 195 °C to 220°C, front and middle were set in the range of 190 °C to 215°C and 180°C to 205°C respectively, rear 2 and rear 1 were in the range of 170 °C to 180 °C and 175 °C to 200 °C respectively. The feeding temperature was set as constant at 50°C with the cooling time of 5 second.

Figure 1: Injection molding machine

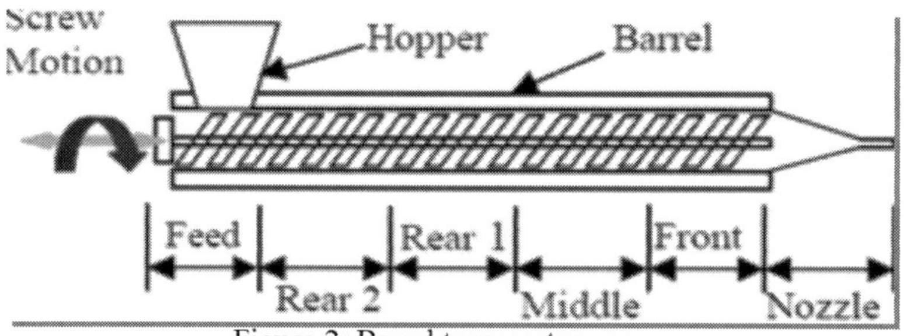

Figure 2: Barrel temperature zone

The influence of composition ratio of volume Cu, nozzle temperature and injection pressure were investigated in this research. Table 3 shows the parameter setting used in this investigation.

Table 3: Investigate Parameter

Factor	Parameter label	Low Level (-1)	Middle	High Level (+1)
Composition ratio volume Cu,%	A	3	5	7
Nozzle temperature,°C	B	195	208	220
Injection pressure, %	C	60	62.5	65

DOE method of two-level full factorial with 4 centre points was implemented to develop the mathematical model for performance predicted of the process. Thus, the research was designed with two-level full factorial where a centre point was assigned for each individual factor to evaluate the curvature trend of the response equations. The investigated responses are tensile strength and thermal conductivity. Standard test specimens were conducted based on DIN EN ISO 527-2 for tensile test and Fourier's Law experimental method for thermal conductivity.

Figure 3 shows the schematic of the experimental set-up for the conductivity test as refer to Sia C.K. et al [6]. The results were calculated using equation 2 with the assumption that all specimens were subjected to the same atmosphere of measurement.

$$Q = - kA\ (T_1\text{-}T_2/L) \tag{2}$$

Where ;

K= thermal conductivity, w/mk
A=cross sectional area, m
T_1-T_2= temperature different, °C
L=thickness,m
Q=rate of heat transfer,joule

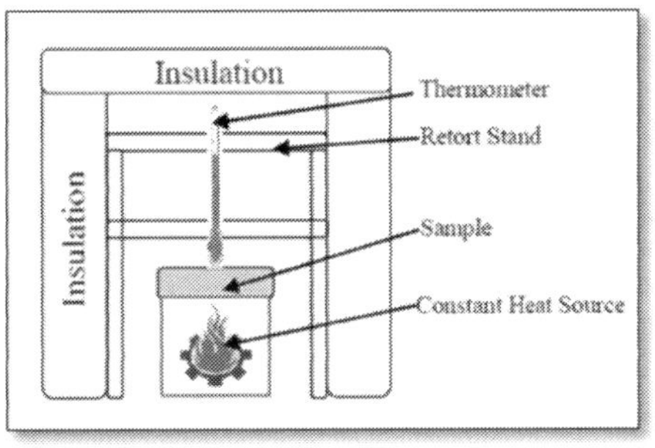

Figure 3: Experimental set-up for thermal conductivity [6]

3. Result and Discussion

All response were statistically analysed through Analysis of Variance (ANOVA) to determine the significant factors including single and the interaction. Total number of 20 experiments were implemented in this research. The revised ANOVA for tensile strength as shown in Figure 4 revealed that the significant single factor are composition of copper, nozzle temperature and injection pressure and the significant interaction factor is the interaction between Cu composition and injection pressure, as indicated by the probability value (p-value) of lower than 0.05.

For Figure 4, the revised ANOVA for thermal conductivity found that the significant single factor is the composition volume ratio of Cu and nozzle temperature while the significant interaction factor is the interaction of Cu composition and nozzle temperature.

ANOVA for selected factorial model

Analysis of variance table [Partial sum of squares - Type III]

Source	Sum of Squares	df	Mean Square	F Value	p-value Prob > F	
Model	248.78	6	41.46	28.30	< 0.0001	significant
A-composition ratio volume Cu	*5.82*	*1*	*5.82*	*3.98*	*0.0676*	
B-nozzel temperature	*37.61*	*1*	*37.61*	*25.67*	*0.0002*	
C-injection pressure	*48.66*	*1*	*48.66*	*33.21*	*< 0.0001*	
AB	*41.31*	*1*	*41.31*	*28.19*	*0.0001*	
AC	*108.27*	*1*	*108.27*	*73.89*	*< 0.0001*	
BC	*7.09*	*1*	*7.09*	*4.84*	*0.0465*	
Residual	19.05	13	1.47			
Lack of Fit	*6.51*	*2*	*3.25*	*2.85*	*0.1003*	*not significant*
Pure Error	*12.54*	*11*	*1.14*			
Cor Total	267.83	19				

Figure 4: ANOVA tensile strength

ANOVA for selected factorial model

Analysis of variance table [Partial sum of squares - Type III]

Source	Sum of Squares	df	Mean Square	F Value	p-value Prob > F	
Model	3.443E-004	3	1.148E-004	8.70	0.0012	significant
A-composition ratio volume Cu	*2.560E-004*	*1*	*2.560E-004*	*19.42*	*0.0004*	
B-nozzel temperature	*1.600E-005*	*1*	*1.600E-005*	*1.21*	*0.2869*	
AB	*7.225E-005*	*1*	*7.225E-005*	*5.48*	*0.0325*	
Residual	2.109E-004	16	1.318E-005			
Lack of Fit	*1.029E-004*	*5*	*2.059E-005*	*2.10*	*0.1422*	*not significant*
Pure Error	*1.080E-004*	*11*	*9.818E-006*			
Cor Total	5.552E-004	19				

Figure 5: ANOVA thermal conductivity

The R square value has also been observed for the source of variation analysis as shown in Figure 6 and Figure 7. R value was found in between 0.6 and 1 indicate that all significant factors were considered in the developed model.

Std. Dev.	1.21	R-Squared	0.9289
Mean	45.98	Adj R-Squared	0.8961
C.V. %	2.63	Pred R-Squared	0.8327
PRESS	44.82	Adeq Precision	20.905

Figure 6: Source of variation analysis for tensile strength

Std. Dev.	3.631E-003	R-Squared	0.6200
Mean	0.11	Adj R-Squared	0.5488
C.V. %	3.25	Pred R-Squared	0.4525
PRESS	3.040E-004	Adeq Precision	7.544

Figure 7: Source of variation analysis for thermal conductivity

Three dimensional surfaces were obtained to illustrate the parameter-response relationship. Figure 8 shows the surface plot graph for tensile strength. It can be seen that the higher the volume of Cu, the lower is the tensile strength which found similar to as found by Mamunya et al. [7]. The optimisation analysis indicated that the low level of Cu composition, low level of nozzle temperature and high level of injection pressure is needed to be used to maximise the tensile strength.

Figure 9 shows the surface plot graph for thermal conductivity. The increasing in the composition of Cu increases the thermal conductivity of the composite material. Although different in parameter range, the response behaviour was found similar as reference [8]. For the thermal conductivity, high level of composition of Cu, low level of nozzle temperature and high level for injection pressure will optimise the response.

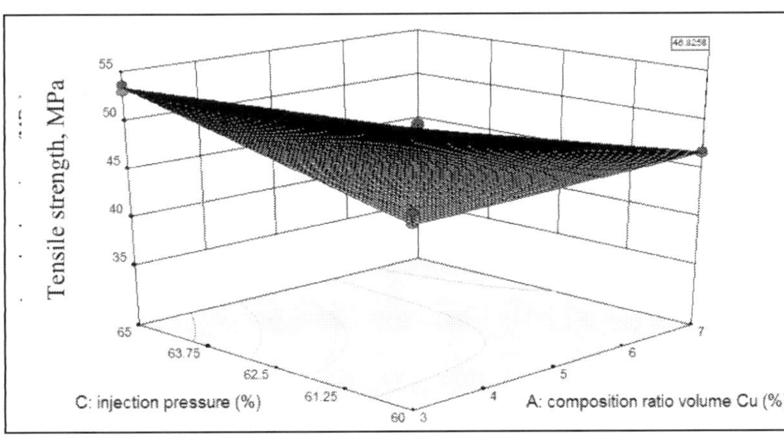

Figure 8: Surface plot graph for tensile strength (σ)

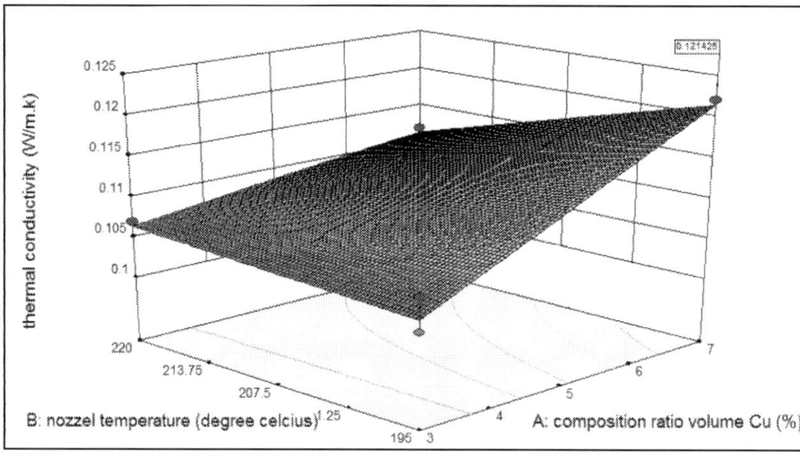

Figure 9: Surface plot graph for thermal conductivity (k)

Mathematical model for each response were developed with 95% confident interval. The predictive model for tensile strenght and thermal conductivity were given by :

Tensile strenght = 209.72566 + 18.87844(A) - 1.77554(B) - 1.12161(C) + 0.064275(AB) - 0.52027(AC) + 0.021304(BC) (3)

Thermal conductivity, k = 0.030213 + 0.019637(A) + 3.45000E-4(B) - 8.50000E-5(AB) (4)

The validity of the developed model was proved through confirmation run. As shows in Table 4, 3 parameter combinations were tested. The result were tabulated (Table 5) and the percentage errors were calculated (Table 6). Results demonstrated that developed model reasonably accurate for the response prediction.

Table 4: Parameter of the confirmation Test

Run	**Composition ratio volume Cu,%**	**Nozzle temperature, °C**	**Injection pressure, %**
1	4	201.25	61.25
2	4	210	63
3	6	215	64

Table 5: Predicted and Actual responses

Run	**Predicted, σ MPa**	**Predicted, (k) W/mk**	**Actual, σ MPa**	**Actual, (k) W/mk**
1	46.0938	0.109769	46.043	0.103
2	46.4496	0.109813	46.606	0.107
3	45.7457	0.112563	45.775	0.111

Table 6: Residual and Error in Prediction

Run	Residual of σ, MPa	Residual of k, W/mk	Error Response 1 (%)	Error Response 2 (%)
1	-.0.0508	-0.006769	0.1	0.6
2	0.1564	-0.002813	0.33	2
3	0.0293	0.0015	0.6	1.35

4. Conclusion

In this research, the relationship between injection molding parameter and the responses were determine experimentally. The mechanical properties of the composite is influenced by certain parameter investigated In this research, the composition ratio volume of Cu is the most significant factor that influence of the mechanical properties of the plastic composite. The best combination parameter setting to maximize the value of tensile strength is the combination of volume percentage of copper powder of 3.00%, the nozzle temperature of 195 °C and the injection pressure of 65%, while the best combination parameter settings to maximize the value of thermal conductivity is the combination of volume percentage of copper powder of 7.00%, the nozzle temperature of 195 °C and the injection pressure of 65%. In overall, the statistical analysis was provided clear understanding on the influence of injection parameter setting through the validated mathematical models.

References

[1] William F. Smith & Javad Hashemi, Ph.D (2011). "Foundation of Material Science and Engineering" Fifth edition (McGraw-Hill), pages 515-517.

[2] R.P. Kusy , in *Metal-Filled Polymers*, edited by S.K. Battacharya (Dekker, New York, 1986)

[3] Raut, S. V, & Khande, R. B. (n.d.). Investigation of the Effect of Mechanical Properties on the Copper filled Acrylonitrile Butadiene Styrene (ABS-Cu Composites).

[4] Sa'de, N.,Masood, S.H., Nikzad, M., Ibrahim, M.H.I. (2013) "International Journal of Engineering Research and Apllication (IJERA) Dynamic Mechanical Properties of Copper-ABS Composite for FDM Feedstock", pp.1257-1263.

[5] M. Nikzard, S, H. Masood, I, Sbarski, Thermo Mechanical Properties of a highly filled Polymeric Composites for fused Deposition Modeling, *Material and Design*, 32, 2011, 3448-3456

[6] Sia, C. K., et al. "Optimum preparation and thermal conductivity of novel glossy paint prepared from waste of palm oil fly ash composite." *International Journal of Automotive & Mechanical Engineering* 12 (2015).

[7] Mamunya, Y.P., Davydenko, V.V., Pissis, P. and Lebedev, E.V. (2002). Electrical and Thermal Conductivity of Polymers Filled with Metal Powders, European Polymer Journal, 38(9): 1887–1897.

[8] Mamunya, Ye P., et al. "Electrical and thermal conductivity of polymers filled with metal powders." European polymer journal 38.9 (2002): 1887-1897.

IOP Conf. Series: Materials Science and Engineering **203** (2017) 012019 doi:10.1088/1757-899X/203/1/012019

Electrical Discharge Machining of Polycrystalline Diamond Using Copper Electrode – Finishing Condition

M.A. Haikal Ahmad[1], M. Zulafif Rahim[1,2,*], M. F. Mohd Fauzi[1], N. Hafizah Azis[1], A.E Ismail[1], Mohd Fahrul Hassan[1], A.M.T.Arifin[1], M.S.Yusof[1,2], M. Rasidi Ibrahim[1]

[1]Faculty of Mechanical and Manufacturing Engineering, University Tun Hussein Onn Malaysia (UTHM) P.O. Box 101, Parit Raja, Batu Pahat 86400, Johor, Malaysia.

[2]Tensei Innovative Solution, No. 51 Block C, Universiti Tun Hussein Onn Malaysia (UTHM) City Campus, Parit Raja, Batu Pahat 86400, Johor, Malaysia.

*zulafif@uthm.edu.my

Abstract. Research on machining process of Polycrystalline Diamond (PCD) is becoming important as the material was believed suitable to be used for cutting tools of advanced aeronautical structure. Electrical Discharge Machining (EDM) was regarded as the suitable method to machine PCD due its noncontact process nature. The objective of this research is to determine the influence of several EDM parameter such as sparking current, pulse duration, and pulse interval to the material removal rate and surface roughness of the machined PCD. Instead of significantly influenced the material removal rate, the sparking current was also highly influenced tha surface roughness. Highest material removal rate of approximately 0.005mm^3/s was recorded by the EDM process with the highest current used of 5A, and lowest pulse interval of 1µs. The influence of pulse duration is not clearly seen at the lowest pulse interval used. On the other hand, 0.4µm was the lowest surface roughness value obtained in this research indicated by the highest sparking current, highest sparking duration and lowest sparking interval of 5A, 1µs and 1µs respectively.

1. Introduction

The interesting discovery of new composite material of CFRP-Titanium stakes brings attention to huge aerospace industries to replace the use conventional CFRP. Though, the main challenge was to machine and fabricate the material due to the extreme hardness of titanium layer. Regarded as among the hardest material in the world, Polycrystalline diamond believed as suitable cutting tools to machine this material. In fact, the application of PCD tools not limited to only for aerospace structure machining process, but also for underwater drilling in oil and gas industries as well as for rock drilling in new outer-space discovery. The investigation on the machinability of PCD would greatly important as it plays important role in several aforementioned extreme applications.

Since 2010 extensive research has been initiated to increase PCD machining efficiency and tool quality [1, 2] But no significant breakthrough was achieved due to the lack of innovative approaches and new theories to solve the two fundamental problems in machining PCD: the ultra-hardness of diamond particles and the low electrical conductivity of PCD material. Despite improvements of various levels were achieved, the increase in machining efficiency was limited. This includes the investigations on using short-pulsed lasers, ultra-short-pulsed lasers and other novel laser-related machining techniques for PCD processing [3, 4]. Precision and near damage-free machining could be achieved by using lasers of ultra-short pulses. The nanosecond-pulsed lasers are likely more desirable for high efficient micromachining where the surface damages can be properly controlled. However, extremely slow process limits its application.

Due to high hardness of PCD, only diamond grinding wheels can be applied in PCD grinding [2]. However, it suffers with high wheel wear (G ratio of 0.01 to 0.05) and extremely low material remove rate [1, 5] This leads to the high manufacturing cost caused by the long machining time that becomes a major impediment to the implementation of PCD tools: a brazed PCD tool is about 10 times expensive than a tungsten carbide tool, and the price of a vined PCD tool is about 20 times higher than the carbide ones

Electrical Discharge Machining (EDM) is a noncontact thermal erosion process in which the metal is removed by a series of recurring electrical discharges between the electrode and the electrically conductive workpiece, in the presence of a dielectric fluid. The electric conductivity of PCD caused by the conductive binding material (Cobalt) makes it possible to machine PCD tools with EDM. However, the low electrical conductivity makes the EDM process is almost 3 times slower than EDM of conductive metals. The fundamental problem that causes low machining efficiency in EDG of PCD is the low electricity conductivity of PCD [6].

This study discusses on the erosion performance of Electrical Discharge Machining (EDM) in PCD machining, specifically in finishing stage. Finishing EDM of PCD considered as a slow process due its high melting temperature of diamond grains (approximately 3400°C) and low electrical conductivity. Therefore, there is a pressing need and powerful incentive to develop a new EDM strategy to significantly increase PCD machining efficiency. As an attempted to discover new machining theories, this experiment was carried out as the benchmark for the hereinafter investigation.

2. Methodology

PCD known as synthetic diamond which is among the hardest materials in the world. It is produced by a sintering process that composed diamond particle and metallic binder of cobalt substance under high pressure and high temperature condition. CTX002 PCD grade was used for this erosion experiment with the dimension of 7mm x 0.5mm for the width and thickness respectively. Copper rod with diameter of 10mm was used as the electrode. Figure 1 shows the PCD and Copper electrode used in this investigation.

Figure 1: a. Polycrystalline diamond (PCD), b. Copper electrode.

Table 1: Properties of CTX002 PCD [7]

PCD Grain Size (μm)	Diamond Fraction (Vol %)	Cobalt Fraction (Vol %)
2	*84.8*	*15.2*

The research was conducted by 4-axis SODICK EDM die sinking machine. Three variable parameter were used in this research; pulse ON, pulse OFF, and current. The process was categorized as high voltage machining as high peak voltage was fixed at 120V for the entire experiment. Negative polarity was used in this investigation as found more suitable for finishing process [8]. The detail machining parameter is shown in Table 2.

Table 2: EDM machining parameter

Parameter	Description
Supply voltage	120V
Current	1A and 5A
Pulse duration (Pulse-on)	1μs and 5μs
Pulse interval (Pulse-off)	1μs and 5μs
Depth of cut	0.1 mm

The erosion performance was indicated by the material removal rate (MRR) and surface roughness (Ra) produced by cooper electrode. Cleaning process using Acetone was done prior to ultrasonic cleaning as it is crucial to remove deposited debris on the machined surface. The roughness measurement was done according to EN ISO 4288 using Surface Roughness Tester Mitutoyo SJ-400. The Material removal rate (MRR) of the work piece was calculated by using the following formula. Where V is volume of material removed and t is the machining time.

$$MRR = \frac{V\ (mm^3)}{time\ of\ machining(t)\ (sec)} = mm^3/sec$$

3. Result and Discussion

Figure 2 shows the MRR obtained by the process where the error bars indicate 95% of confident interval. It can be seen that the higher the sparking current is, the faster is the erosion process. As shown in Figure 2a, the MRR obtained by 5A was recorded approximately 800% higher than the MRR obtained by the sparking current of 1A. Similar to as found by the EDM of ordinary material, it was due to the production of higher sparking energy. At lower interval, the influence of pulse duration is not significant. Although the mean value of MRR is higher for small pulse duration, 95% confident interval indicates there is a possibility to obtained similar MRR compared to 5μs pulse interval.

At high sparking current of 5A, the influence of pulse interval can be clearly seen. This can be due to several reasons: (1) Due to high thermal conductivity of PCD, fast heat dissipation can be expected. Hence, increasing pulse interval (pulse-off) from 1μs to 5μs would significantly increase the probability of heat losses before the next sparking event, (2) The interval time between spark isn't enough for dielectric strength recovery. Hence, immature sparking frequently happened which resulted to inefficient sparking process. Due to the static electrode implemented, slow dielectric flow under the gaps can be expected which made slower strength discovery. As reported by reference [9], faster the electrode rotation is, the faster is the dielectric strength recovery and removal process.

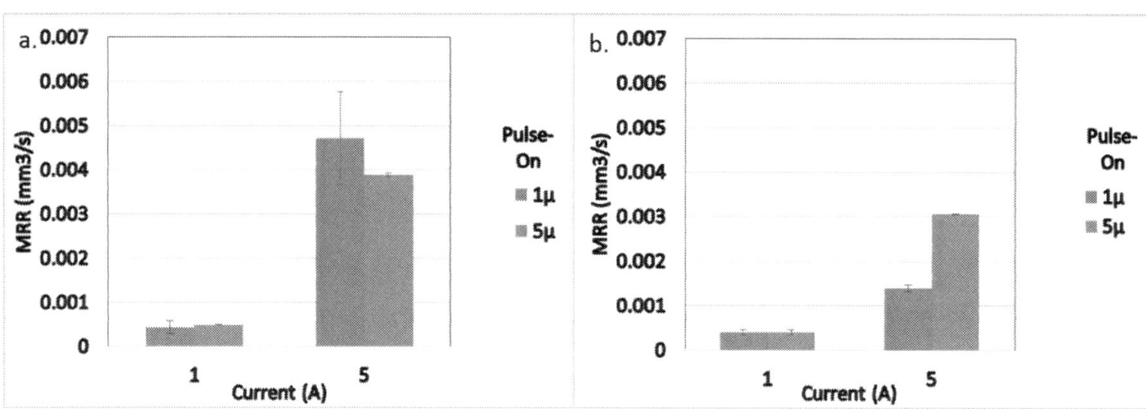

Figure 2: Result on Material Removal Rate (MRR); a. Using 1μs Pulse Interval (Pulse-Off) b. Using 5μs Pulse Interval (Pulse-Off)

In general, the process produced the surface roughness of lower than 1μm. The value was higher than some publised data of PCD EDM with rotating electrode in reference [10]. The higher in surface roughness value was due to inefficient debris removal in EDM of static electrode. The debris accumulation caused to the spark concentrion, consequently resulted in worse surface roughness [9-11].

Interestingly, this investigation revealed that the lower current is not necessarily able to produce lower surface roughness. As shown in Figure 3, lower surface roughness was produced by the EDM with 5A sparking current that is normally not the case for ordinary material. In EDM of ordinary material, higher peak current of the same pulse duration reported as produced higher sparking energy. This resulted in production of deeper crater and rougher surface [12].

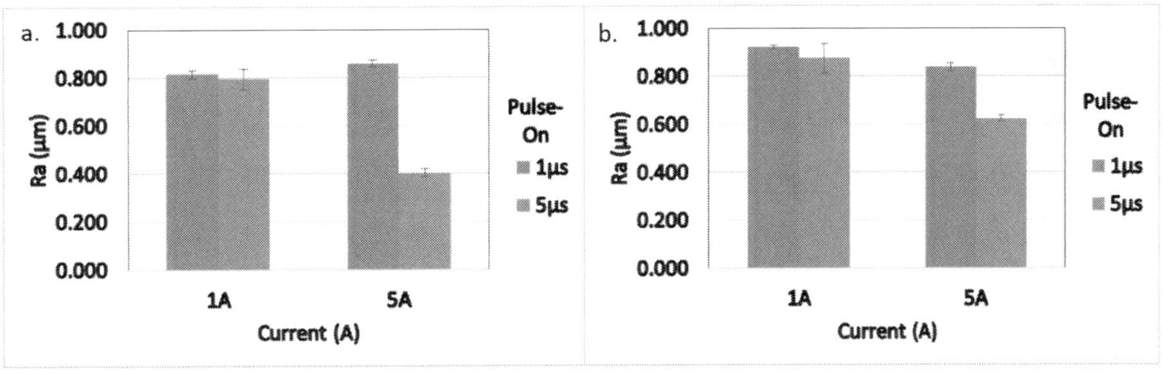

Figure 3: Result on Surface Roughness (Ra); a. Using 1µs Pulse Interval (Pulse-Off) b. Using 5µs Pulse Interval (Pulse-Off)

As discussed in reference [13], PCD EDM undergo a unique removal mechanism. Breaking mechanism of diamond grains reported as the erosion mechanism for roughing while graphitisation erosion for finishing process [13]. If this is the case, better surface roughness obtained by 5A could be due to the suffient graphitisation activation energy provided by the process.

The researchers believed that clear decription on the efficient erosion in production of smooth PCD surface can be explained by carefully analyse the sparking profile produced by the process. As mentioned by The outcome would then elucidate the erosion mechanism of PCD that believed distinctive from the ordinary material.

4. Conclusion

It was theorized that the material removal rate and surface roughness is closely related to the sparking energy produced by the process. In the case of material removal rate, the high current used of 5A was produced bigger sparking energy, thus made faster to the erosion process. In this study, 0.0047 mm/s recorded as the highest MRR. 5A current was also influenced to the production of lowest surface roughness. Yet, the reduction of surface roughness with the higher sparking energy remained undiscovered. To reveal the exact reason, the observation of the spark profile is essentially required that would be the future extension of this research.

Acknowledgement

The authors would like to acknowledge the Ministry of Higher Education (MOHE) Malaysia under Fundamental Research Grant Scheme (FRGS) grant, Vot 1586 and Universiti Tun Hussein Onn Malaysia (UTHM) under Postgraduate Research Grant (GPPS), Vot U805 for financial support.

References

1. Iwai, M., S. Ninomiya, and K. Suzuki, *Effect of complex electrodischarge grinding for electrically conductive PCD.* Advanced Materials Research, 2011. **325**: p. 276-281.
2. Qinjian, Z., et al., *Study on electrical discharge and ultrasonic assisted mechanical combined machining of polycrystalline diamond.* Procedia CIRP, 2013. **6**: p. 589-593.
3. Wu, Q. and J. Wang. *An experimental investigation of the laser milling process for polycrystalline diamonds.* in *Advanced Materials Research.* 2011. Trans Tech Publ.
4. Perry, M., et al., *Ultrashort-pulse laser machining of dielectric materials.* Journal of Applied Physics, 1999. **85**(9): p. 6803-6810.
5. Liu, G., et al. *Study on grinding technology of electrical discharge grinding of curve edge polycrystalline diamond.* in *Key Engineering Materials.* 2006. Trans Tech Publ.
6. Li, G., et al., *Experimental study on quality of PCD tools machined by different electric discharge grinding processes.* Cogent Engineering, 2016. **3**(1): p. 1228234.
7. Rahima, M.Z., et al., *ELECTRICAL DISCHARGE GRINDING VERSUS ABRASIVE GRINDING IN POLYCRYSTALLINE DIAMOND MACHINING.* 2015.
8. Rahim, M.Z., S.L. Ding, and J. Mo. *Electrical Discharge Grinding (EDG) of Polycrystalline Diamond-Effect of Machining Polarity.* in *Advanced Materials Research.* 2014. Trans Tech Publ.
9. Hu, B., et al. *Experimental Study of Wheel Rotating Speed Effect on Electrical Discharge Grinding.* in *Applied Mechanics and Materials.* 2015. Trans Tech Publ.
10. Rahim, M., et al., *Electrical discharge grinding versus abrasive grinding in polycrystalline diamond machining—tool quality and performance analysis.* The International Journal of Advanced Manufacturing Technology, 2015: p. 1-15.
11. Rahim, M.Z., S. Ding, and J. Mo, *Electrical discharge grinding of polycrystalline diamond–Effect of wheel rotation.* Machining science and technology, 2016. **20**(1): p. 62-78.
12. Han, F., J. Jiang, and D. Yu, *Influence of discharge current on machined surfaces by thermo-analysis in finish cut of WEDM.* International Journal of Machine Tools and Manufacture, 2007. **47**(7–8): p. 1187-1196.
13. Zulafif Rahim, M., S. Ding, and J. Mo, *Electrical Discharge Grinding of Polycrystalline Diamond—Effect of Machining Parameters and Finishing In-Feed.* Journal of Manufacturing Science and Engineering, 2015. **137**(2): p. 021017-021017.

Improving Educator Development by Innovation in Teaching Activity via web 2.0

Nurhanim Saadah Abdullah, Mohd Ismail Abd Aziz, *Affero Ismail, and Suhaizal Hashim

Universiti Tun Hussein Onn Malaysia

Faculty of Technical and Vocational Education

86400 Parit Raja, Batu Pahat, Johor, Malaysia.

*affero@uthm.edu.my

Abstract. Preparing insightful teaching and learning materials for a lesson does need the effort from the educators. Educators should make some research of suitable ways to improve their teaching and learning sessions. In this 21st century, technologies are widely used as tools for education. Even so, there are educators that willing to support and some who do not agree to change. The aim of this study is to develop an innovation teaching materials by applying web 2.0 tools. The intention is to broaden knowledge and in the same time getting response and feedback from people regarding the teaching and learning session materials produced with proper instruction. Action research was used to give a structured flow of this study. The outcome of this study was encouraging and the reflection of this study can help educators in improvising their teaching and learning sessions and materials using action research.

1. Introduction

Being educators is a big job and responsibilities are on the educators shoulder especially to educate multiple generations such as Baby boomers, generation X and generation Y with their own differences in characteristics. The educations now are moving towards 21st century and educators need to move faster and prepare their lessons based on 21st century education. A student need electronic tool to support students' learning and professional growth [1]. Therefore, conducting learning session is more feasible to student's current needs. Web 2.0 is used in this study as it emphasize user-generated content, usability and interoperability for end users. Meanwhile Web 1.0 refers to the first stage in the World Wide Web, which was entirely made up of Web pages connected by hyperlinks. Although the exact definition of Web

1.0 is a source of debate, it is generally believed to refer to the Web when it was a set of static websites that were not yet providing interactive content. In Web 1.0, applications were also generally proprietary.

2. Objective and Methodology

This study is focusing on the development of teaching material and applying web 2.0. Action research was implemented in which participants examine their own educational practice systematically and carefully, using the techniques of research. It is based on the following assumptions [2]:

- Teachers and principals work best on problems they have identified for themselves
- Teachers and principals become more effective when encouraged to examine and assess their own work and then consider ways of working differently
- Teachers and principals help each other by working collaboratively
- Working with colleagues helps teachers and principals in their professional development

According to Ferrance [3], action research specifically refers to a disciplined inquiry done by a teacher with the intent that the research will inform and change his or her practices in the future.

3. Past Paradigms and Anomalies that occurred in Instructional Technology

Paradigm in education keeps shifting until recent years and Instructional Technology in education is one of the important matters that have experienced various paradigm shifts [4]. Technologies are being used in order to support instruction. The popular technology used in instruction is of course the computers. Before computers, there were numbers of other forms of technology used in instruction such as radio, television and film been introduced into classroom [5].

Anomaly that occurs at that particular era was as described by Cuban [5] which he argued that the failure of various technology-driven initiatives to achieve an appreciable impact has been due largely to a failure on the part of the designers to fully appreciate the expectations and requirements of classroom practitioners. Lack of appreciation of classroom practitioners has caused a failure to a technology in instruction. Therefore, in the next segment, the researcher would like to describe more on anomalies in the use of instructional technology focuses on higher education.

The rationale for formulating 21st century skills has been summarized in the table based on Dede's [6] comparison between 20th century & 21st century skills. From Table 1, it is obvious that the difference between 20th century and 21st century skills is at the usage of technology. The knowledge communicates in schools for 21st century skill is towards creating new contextual skills that suitable for 21st century learning environment.

Table 1. Summary of the difference of 20th century and 21st century

	20th century skills	21st century skills
Capabilities people need for work	✓	✓
Citizenship	✓	✓
Self-actualization	✓	✓
Emergence of every sophisticated information and communication technologies		✓ eg: types of work done by people-as the kinds of labor done by machines- shifted as computers and telecommunications capabilities can accomplish human tasks
Knowledge communicated in schools		✓ Creating new "contextual" skills unique to millennium work and citizenship

4. Results and Discussion

There are four basic themes in action research that are empowerment of participants, collaboration through participation, acquisition of knowledge, and social change [3]. In conducting action research, educator structure routines for continuous confrontation with data on the health of a school community. These routines are loosely guided by movement through five phases of inquiry. The five phrases of inquiry are simplified to better understanding as in Figure 1, the action research cycle.

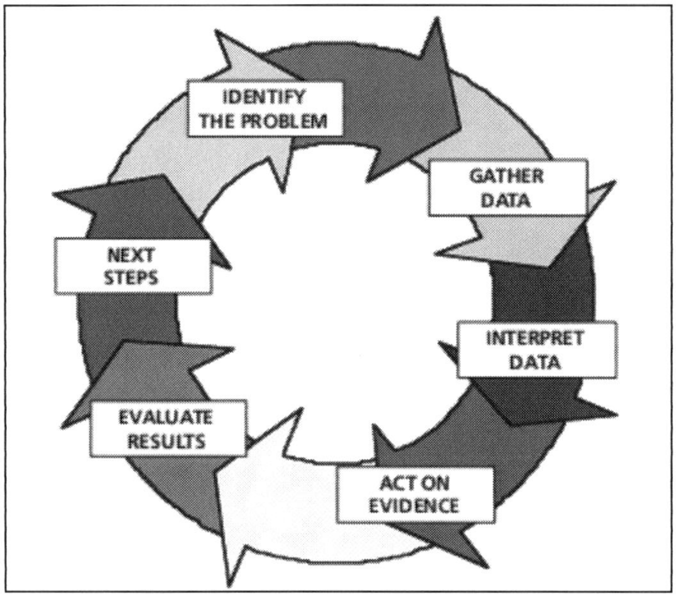

Figure 1. Action Research Cycle [3]

To make the class lively, the instructor must prepare activities that can attract students and entice them to actively continue to learn. According to Swanson, Cano, Samy, and Hynes [7], the use of the term "active learning" by educators relied more on intuitive understanding than a common definition. Thus, many educators claim that all learning is essentially active and students are actively involved listening to formal presentations in the classroom [7]. Students must do more than just listen [8]. They must read, write, discuss, or be engaged in solving problems. Therefore, strategies that promoting active learning be defined as instructional activities involving students doing things and thinking about what they are doing [8]. Therefore, educators must be creative and advance in creating a better teaching and learning session to make learners more engaged and gain meaningful lessons.

After identifying all the problems stated in previous section, the researcher has done an action research to improve teaching and learning session in vocational education. The researcher has the idea of upgrading the traditional lesson to using web 2.0 tools to expose the lesson to other learners not specifically to vocational learners and instructors only but to other audience.

4.1 Preparation of Learning Material

In this study, a lesson of cooking was selected to be improved in vocational education. The old method that still being applied today is instructor do the demonstration to the students and students try to follow the steps. At the end of the class, the objective to make the cake has been reached but only to that extent. The researcher wanted to broaden the classes to be online and can be watched by other people too. The idea is by recording the lesson, and publish it online using Web 2.0.

The benefit of using web 2.0 is to get feedback from the viewer where they can post their comments and we can reply the comments. These threads of constructive comments are used as reflection of the lesson itself and how the lesson can be improved next time. As this activity will involve instructor and learners, a suitable theory has been chosen. Activity Theory is chosen because the intention of the researcher is to create a social constructivist learning environment. Therefore, activity theory is suitable to be applied in this study.

In the work of Lorna Uden and Valderas [9], they explain that activity theory focuses on the interaction of human activity and consciousness within its relevant environmental context. The basic structure of an activity for this study can be illustrated as in Figure 2.

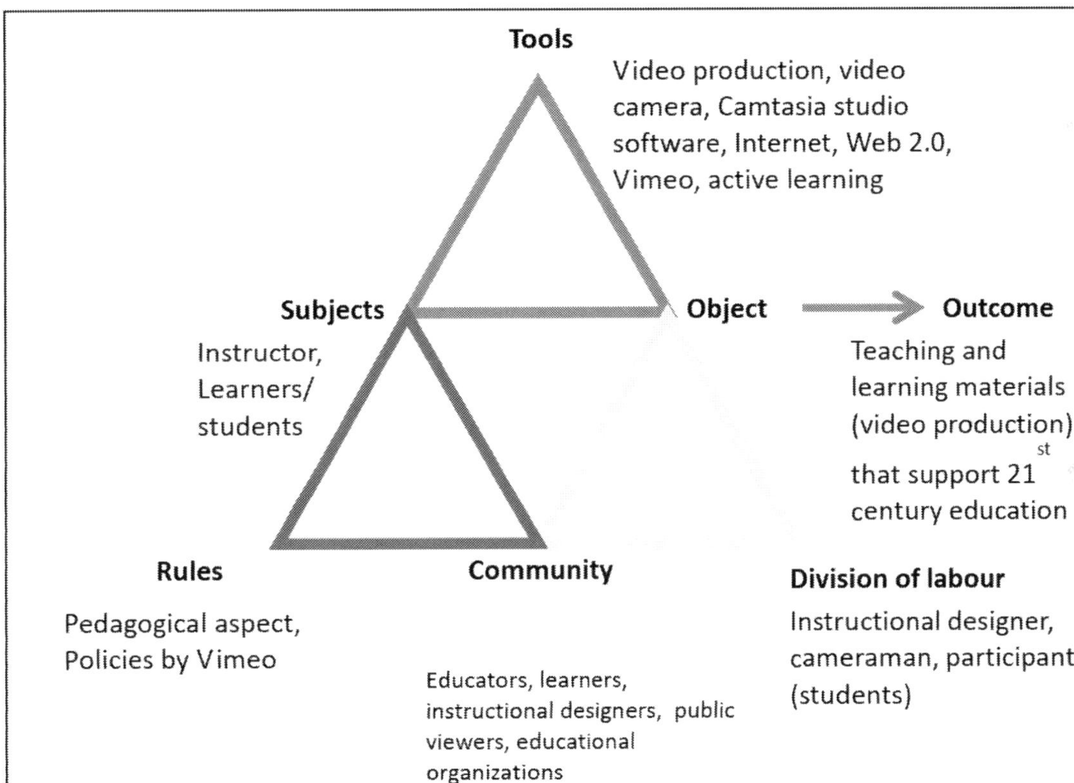

Figure 2. Activity system for peer review assessment video

The researcher used almost all tools in the while editing the videos such as cutting the video length, adjusting the audio (high or low), put captions for easy and clear tracking by using callout tools, use title clips for every intersections of the videos, use transitions and basic video editing functions such as blurring and fading the video.

4.2 Publishing into web 2.0

After the video production processes were done, it is time to publish the videos to any online video streaming such as You Tube and Vimeo video sharing. The researcher preferred Vimeo since the layout is more interesting and colorful. Besides, the user friendly environment offered by Vimeo has attracted the researcher's interest to upload the videos into Vimeo. In order to upload the videos, the researcher need to sign up an account and it is free, and Vimeo has given 500 megabyte of space for users to upload their videos. The important reminder from Vimeo is that the videos uploaded must be original from the users. The process of uploading the videos depends on the length of the videos itself. When the uploading process between user's computer and Vimeo are done, the user need to wait for another approximately 33 minutes for the videos to be posted on Vimeo and can be viewed online (Figure 3 & 4).

Figure 3. Video editing process (post-production) using Camtasia Studio software

Figure 4. Video on Vimeo online video sharing

4.3 Review of the video on Vimeo

User can send invitation to other colleagues and students and also other audience to view the videos by entering email addresses of the person. Vimeo will automatically send invitation emails to the specified person. If the user set the videos as private, then password are required for other person to view the videos. As for the researcher, the videos are set to Public so that not only specified person can view, but other viewers can also view the videos without needing to enter password. The intention is to share the knowledge with the world as a big community.

The real intention of producing the videos was to get feedbacks and comments from colleagues and students who viewed the videos. Viewers can interact with the user by posting their comments in the comment box. Users can reply each of the comments by easily clicking the 'reply' button on the left of the comment box post by the reviewers.

The process of posting comments and giving feedbacks are the most important process in gathering information of the current method used by the educator. From the comments, the instructor can find out the strength and weaknesses from the lesson. Therefore, after identifying all the strength and weaknesses from the video, the educator can make a self-reflection in order to improve the next lesson. As for the researcher's videos, positive comments and ideas were posted by the viewers and those comments surely reflect the teaching and learning process occurred in the videos. The researcher was excited to get more and more comments and feedback from the other viewers so that amendments can be done in the next lessons.

From the thread of comments in Vimeo, it shows that viewers are actively concentrating on the videos. There were positive comments and also some viewers posted ideas regarding the improvement that can be done for the next video production. Researcher has had a great experience in doing the whole process starting from planning until the reflection and improvement to be done. Some comments that can improve the video productions for educational purposes. Some of the comments are as follows:

"The instructor used authentic learning by having the training of preparing a cake in a kitchen. The introduction is clear and it was confidently and naturally presented. The whole video presentation looks professional. You could also use some camera angle techniques to improve the presentation. I looks like instructor is not only teaching the people she has invited but also the audience who is watching the video. Good work."

"Student gave innovative suggestion to replace the cake batik with other ingredient, instructor didn't encourage new innovative and creativity; even though instructor explain why other ingredient was not suitable."

The educator will definitely list the weaknesses with the purpose of making improvements in self teaching and learning session. Important aspect as pedagogy in teaching cannot be left as second priority. All these processes are actually a good approach of learning and embedding technologies in daily lessons as experienced by the researcher. The combinations of ideas to improve teaching and learning session using technology in order to promote 21st century education can be seen positive by the application of action research done by the researcher. What educators should do is find ways to always improve the current practice in teaching and learning so that the meaningful learning can occur. Students in the 21st century are active and full of commitments. Therefore, it is the role of educators to be ahead of the students to give motivation and engagement to the learners towards knowledge taught. Educators also need to look in broader view. Educator needs to put more effort in sharing their knowledge to produce quality graduates [10].

5. Conclusion

To conclude, each of the educators in 21st century should be aware of the ICT transition. Students are now attracted to interactive contents and mobile learning. This effort could safe cost, time and the knowledge could be shared to the entire technology world [11]. The development of online material does support the 21st century education and also enhance the educator's skill in using web 2.0 tools in making learning session meaningful with proper instructional technologies [12].

References

[1] Salonen, P. (2000). *Development project of clinical training 2004-2006*, Final report, Centria University of Applied Sciences.

[2] Watts, H. (1985). When Teachers Are Researchers, Teaching Improves, *J. Staff Dev.*, vol. 6, no. 2, pp. 118–127.

[3] Ferrance, E. (2000). *Action research*. LAB, Northeast and Island Regional Education Laboratory at Brown University.

[4] Koschmann, T. (1996). Paradigm shifts and instructional technology: An introduction, *CSCL Theory Pract. an Emerg.*

[5] Cuban, L. (1986). *Teachers and machines: The classroom use of technology since 1920.*

[6] Dede, C. (2010). Comparing frameworks for 21st century skills, *21St Century Ski.*, pp. 51–76.

[7] Swanson, B. Cano, J., Samy, M. and Hynes, J. (2007). Introducing active teaching–learning methods and materials into Egyptian agricultural technical secondary schools, *Proc. 23rd.*

[8] Chickering, A. W. and Gamson, Z. F. (1987). Seven principles for good practice in undergraduate education, *AAHE Bull.*, vol. Mar, pp. 3–7.

[9] Lorna Uden, O. P. and Valderas, P. (2008). An activity-theory-based model to analyse Web app, Professor T.D. Wilson, 2008.

[10] Abiddin, N.Z. and Ismail, A. (2014). Exploring Service and Support Needs in Postgraduate Education towards the Higher Education Quality, *Asian Social Science,* 10 (17), 52

[11] Strategic Plan For Online Education (2016). The Cost Of Online Education . The Innovation And Online Committee Florida Board of Governors, Affordability Workgroup 2025.

[12] Lawrence S. Bacow, William G. Bowen, Kevin M. Guthrie, Kelly A. Lack and Matthew P. Lon (2012). Online Learning Systems in U.S. Higher Education. United State: ITHAKA.`

IOP Conf. Series: Materials Science and Engineering **203** (2017) 012021 doi:10.1088/1757-899X/203/1/012021

Vibration analysis of circular plates in contact with fluid: A numerical approach

Anju V Nair[1], Abdul Rahman Mohd Kasim[1*], Mohd Zuki Salleh[1]

[1]Faculty of Industrial Sciences and Technology, Universiti Malaysia Pahang

Lebuhraya Tun Razak, 26300, Kuantan, Malaysia

Corresponding author: *rahmanmohd@ump.edu.my*

Abstract. A mathematical formulation to obtain the non-dimensionalized added virtual mass incremental factor for uniform circular plate in contact with fluid is presented. Based on Rayleigh's quotient, the natural frequencies can be evaluated using the added virtual mass incremental factors. A suitable approximation is used to describe the effect of non-dimensionalized added virtual mass incremental factors in the vibration analysis of circular plates. The numerical results obtained from the present formulation are compared with the known results reported in the literature and a good agreement was found. The maximum error percentage has been predicted from the present formulation is 0.005%.

1. Introduction

Numerical analysis of fluid structure interaction problems have wide range of applications in engineering field. The interaction problems can be considered in the signalling problem of submarines, vibration analysis of ship structures, dams, nuclear reactors and so on. The prediction of natural frequency changes is significant for designing structures in contact with fluid because the natural frequencies in fluid are different from those in air. The numerical approaches such as finite element method as well as boundary element method can handle complex boundary conditions of structures and fluids.

Circular plates in contact with fluid have recently been studied. The vibration analysis of circular plates in contact with fluid has been investigated by Kwak [1]. Based on Fourier - Bessel series approach, the non-dimensionalized added virtual mass factor is evaluated using integral transformation technique. It is also studied the influence of fluid on the natural frequencies in this work. Jeong [2] studied the hydroelastic vibration of two identical circular plates coupled with a bounded fluid. An analytical method based on Fourier - Bessel series approach and Rayleigh Ritz method is used in this work. By using Navier – Stoke's equations, finite element analysis of fluid flows coupled with structural interactions has been described in [3]. Large displacement fluid structure interaction problems has been solved by Heil [4] using Newton's method. It is also reported the importance of consistent stabilisation in the numerical simulation of fluid structure interaction problems. Kwak and Kim [5] investigated the axisymmetric vibration of circular plates in contact with fluid. The research is performed to study the effects of fluid on the natural frequencies by evaluating the added virtual mass incremental factor using integral equation method. The natural frequencies of circular plates are evaluated by using added virtual mass incremental factor. Amabili and Kwak [6] considered the free vibration analysis of circular plates in contact with liquids by revising the Lamb problem. The natural frequencies of the circular plate are calculated for both simply supported and clamped boundary conditions. Based on perturbation technique and the Henkel transformation method, the effect of surface waves on free vibrations of circular plates resting on a fluid surface has been presented in [7]. The frequency and mode shape parameters for plates in vacuo are evaluated in this study. Moreover, the frequency parameters for circular plates having free edge, simply supported and clamped boundary conditions are solved by using characteristic equations. The fluid structure interaction problem with vibration of circular plates for varying thickness has been analysed by Liu and Chang [8]. Based on non-linear properties of structures, the fluid structure interaction problem has been solved in [9]. A combined finite element/ boundary element method for fluid structure interaction problems has been presented by Opstal et.al. [10]. The nonlinear flutter behavior of orthotropic composite rectangular plates under aerodynamic pressures and transverse excitations has been presented by Chen and Li[11]. The Hamilton's principle is employed to derive the nonlinear governing equations of motion based on the third-order shear deformation theory and the von-Karman nonlinear strain–displacement relationship. The Galerkin method is utilized to discretize the equations of motion to a set of nonlinear ordinary differential equations. Numerical simulations are conducted to investigate the futter characteristics and nonlinear dynamics of the composite laminated plate. Shao et.al [12] studied the flutter and thermal buckling behavior of laminated composite panels embedded with shape memory alloy (SMA) wires. The classical plate theory and nonlinear von-Karman strain-displacement relation are employed to investigate the aero elastic behavior of the smart laminated panel. The effects of ply angle of the composite panel, SMA layer location and orientation, SMA wires temperature, volume fraction and prestrain on the buckling, flutter boundary, and amplitude of limit cycle oscillation of the panel are analyzed in this research. The stability and nonlinear vibration of a composite laminated plate with simply supported boundary conditions in subsonic compressible airflow subjected to transverse periodic external excitation are investigated by Yao and Li [13]. The equation of motion of the plate is established by using the von Karman's nonlinear plate theory. The linear potential flow theory considering the compressibility of the airflow is adopted to formulate the aerodynamic model. The critical divergence velocity and the flutter velocity of the plate are obtained by analyzing the natural frequencies of the linear system.

The numerical analysis procedures are based on numerous mathematical computations and are highly time consuming. This makes the designers to search for alternative solution procedures with less mathematical complexity and computational efforts. In the present study, the vibration analysis of circular plates in contact with fluid is studied by using suitable approximations and the added virtual mass incremental factors for uniform thickness circular plate are evaluated. The use of numerical formulation can minimise the amount of time spent using experimental techniques to find many design alternatives. It will help in solving large industrial and mechanical problems.

2. Mathematical formulation

Consider a circular plate in contact with fluid of radius '*a*' and thickness '*h*'. Let 'F' denotes the fluid domain, 'S_a' denotes the surface between the fluid and rigid wall, 'S_b' denotes the surface between the fluid and the plate, and 'S_∞' denotes the surface at infinity respectively.

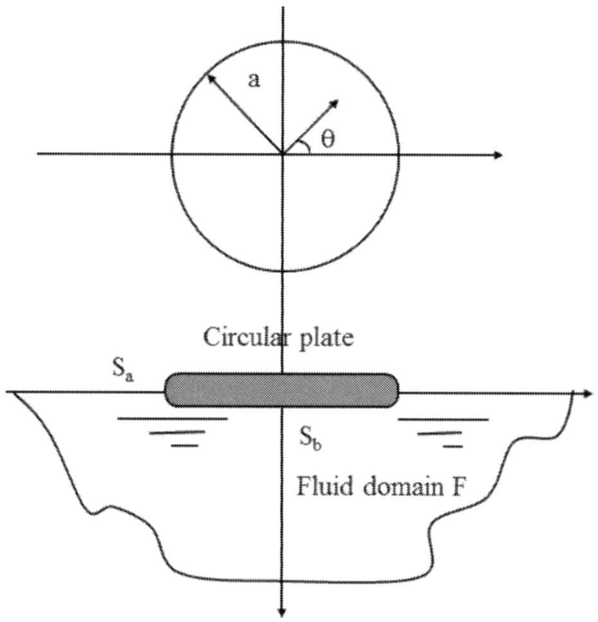

Figure 1. A circular plate in contact with fluid

The differential equation of motion for the circular plate in contact with fluid can be written as

$$D\nabla^4 w+\rho_p \frac{\partial^2 w}{\partial t^2}=0 \qquad (1)$$

where w is the deflection of the plate, $D=\dfrac{Eh^3}{12\left(1-\nu^2\right)}$ is the flexural rigidity, E is the Young's modulus, ρ_p is the mass density of the circular plate and $\nu = 0.3$ is the Poisson's ratio.

Suppose that the contacting fluid is incompressible, inviscid and irrotational. Because of the irrotational behaviour of the fluid, there should be a velocity potential, which can be represented as

$$U(r,\theta,z,t)=\phi(r,\theta,z)T(t)=\phi(r,z)T(t) \tag{2}$$

where the spatial velocity potential $\phi(r,z)$ satisfies the Laplace equation

$$\nabla^2\phi=\frac{\partial^2\phi}{\partial r^2}+\frac{1}{r}\frac{\partial\phi}{\partial r}+\frac{\partial^2\phi}{\partial z^2}=0 \text{ in fluid domain F} \tag{3}$$

The condition of the rigid wall on S_a can be written as

$$\left.\frac{\partial\phi(r,z)}{\partial z}\right|_{z=0}=0 \quad \text{on } S_a \tag{4}$$

Furthermore, the interaction between the fluid and the plate can be represented as

$$\left.\frac{\partial\phi(r,z)}{\partial n}\right|_{z=0}=-W(r) \text{ on } S_b \tag{5}$$

where $W(r)$ is the mode shape of the plate vibrating in contact with the fluid.

For a free surface,

$$\phi,\frac{\partial\phi}{\partial r},\frac{\partial\phi}{\partial z}\rightarrow 0 \text{ as } r,z\rightarrow\infty \text{ on } S_\infty \tag{6}$$

Using Rayleigh's quotient, we can write

$$f_a^{\,2}\alpha\left(\frac{V_p}{T_p^*}\right)_{air} \text{ and } f_l^{\,2}\alpha\left(\frac{V_p}{T_p^*}+T_l^*\right)_{fluid} \tag{7}$$

where f_a is the natural frequency of the plate, f_l is the natural frequency of the plate in contact with the fluid, T_p^* and V_p are the reference kinetic energy and maximum potential energy of the circular plate and T_l^* is the reference kinetic energy of the fluid due to the motion of the plate.

The relation between reference ad maximum kinetic energies can be written as

$$T_{\max}=T^*\omega^2 \tag{8}$$

where ω is the frequency in radians per second. The expression (8) is adapted from [14].

Based on the assumption that the dynamic loading of fluid has an insignificant effect on mode shapes, the following equation can be obtained from eqn (7).

$$f_l = \frac{f_a}{\sqrt{1+\beta}} \tag{9}$$

Here, β is called the added virtual mass incremental factor (AVMI factor) can be defined as the ratio of the kinetic energy of the plate due to the motion of the plate over the kinetic energy of the circular plate itself.

Therefore, β can be written as

$$\beta = \frac{T_l^*}{T_p^*} = \Gamma \frac{\rho_l}{\rho_p}\left(\frac{a}{h}\right) \tag{10}$$

where Γ is called the non-dimensionalized added virtual mass incremental factor (NAVMI factor), ρ_l is the density of the fluid and ρ_p is the mass density of the circular plate.

The reference kinetic energy due to the presence of fluid can be evaluated by the velocity potential as

$$T_l^* = -\frac{1}{2}\rho_l \int_0^{2\pi}\int_{-\infty}^{\infty} \frac{\partial\phi(r,\theta)}{\partial n}\phi(r,0)\,r\,dr\,d\theta \tag{11}$$

The reference kinetic energy of the plate can be written as

$$T_p^* = \frac{1}{2}\rho_p \int_0^{2\pi}\int_0^{a} hW^2(r)\,r\,dr\,d\theta \tag{12}$$

where $W(r)$ is the mode shape of the plate.

Suppose that the variable thickness is of the form

$$h(r) = h_0 f(r)\ ;\ f(r) = 1-\mu r^2 \tag{13}$$

where h_0 is the thickness of the center of the plate and μ is the taper parameter of the varying curve.

Besides, suppose that

$$w(r,t) = W(r)T(t) = W(r)e^{i\omega t} \tag{14}$$

Therefore, the NAVMI factor can be evaluated from eqn (10) by calculating T_l^* and T_p^* and f_l can be evaluated from eqn (9). By using the NAVMI factor, the natural frequencies can be evaluated.

IOP Conf. Series: Materials Science and Engineering **203** (2017) 012021 doi:10.1088/1757-899X/203/1/012021

3. Numerical results and discussions

The present study investigates the NAVMI factor of uniform thickness circular plate in contact with fluid. The equations are solved by using a suitable admissible function with boundary conditions. The following admissible function is considered in the present study.

$$F = b_0\left(1-\left(\frac{r}{a}\right)^2\right)^n$$

It should satisfy the boundary conditions.

At r = 0, $\frac{\partial w}{\partial r} = 0$; At r = *a*, *w* = 0, $\frac{\partial w}{\partial r} = 0$

In this study, the following parameters for uniform circular plate are used.

a = 1m; h_0 = 0.5m; $\rho_p = 2.44\times10^3\ kg/m^3$, $\rho_l = 1000 kg/m^3$

The following table represent the values of NAVMI factors for a uniform thickness circular plate. In this table, '*n*' represents the order of the mode number and '*N*'denotes the number of terms used in the interpolation function to approximate the mode shape. The present numerical results are compared with the numerical results obtained from the known literature. It can be say that the results match well with the known results.

Table1: Non-dimensionalized added virtual mass incremental factor for uniform circular plate in contact with fluid.

N/*n*	1	2	3	4	5
1	0.6830 (0.6826)*	-	-	-	-
2	0.6675 (0.6671)*	0.3155 (0.3153)*	-	-	-
3	0.6673 (0.6668)*	0.2833 (0.2829)*	0.2071 (0.2068)*	-	-
4	0.6670 (0.6668)*	0.2812 (0.2808)*	0.1754 (0.1752)*	0.1535 (0.1530)*	-
5	0.6669 (0.6668)*	0.2806 (0.2800)*	0.1676 (0.1670)*	0.1294 (0.1291)*	0.1198 (0.1192)*

* denotes the values from the known literature [8].

From Table 1, it can be shown that the values of the NAVMI factor decreases with respect to order of the mode numbers. It is clear that the presence of fluid effects the values of NAVMI factor. Moreover, the existence of fluid has a significant impact on the mode of the circular plate. Also, it can be noted that the values of the NAVMI factor in the first mode has higher values compared to the other modes.

Table 2: The percentage error of NAVMI factor for uniform circular plate in contact with fluid to the known results.

N/*n*	1	2	3	4	5
1	0.0005	-	-	-	-
2	0.0006	0.0006	-	-	-
3	0.0007	0.0014	0.0015	-	-
4	0.0003	0.0014	0.0011	0.0033	-
5	0.0001	0.0021	0.0036	0.0023	0.0050

Table 2 describes the percentage error of NAVMI factor for uniform circular plates in contact with fluid to the known literature results. The highest percentage error obtained from reference is 0.0050%. Hence it can be say from the tables that the results from the present investigation are in good agreement with those obtained from known literature.

4. Conclusion

A numerical study has been carried out to obtain the non-dimensionalized added virtual mass incremental factor (NAVMI factor). A suitable approximation with the help of an admissible function is used in the present study to analyse the influence of fluid on the circular plate. The numerical values of NAVMI factor is evaluated for various mode numbers and are compared with the known results. The present numerical results match well with the known literature values. Also, it is clear that the presence of fluid effects the values of NAVMI factor.

Acknowledgement
The authors would like to thank the Universiti Malaysia Pahang and Ministry of Higher Education, Malaysia for the provision in terms of financial form from the research grants (RDU160330).

Nomenclature

a	= Radius of the plate
h	= Thickness of the plate
F	= Fluid domain
S_a	= The surface between fluid and rigid wall
S_b	= The surface between fluid and plate

S_∞ = The surface at infinity

w = Deflection of the plate

D = Flexural rigidity

E = Young's modulus

ρ_p = Mass density of the plate

ρ_l = Density of the fluid

ν = Poisson's ratio

$U(r, \theta, z, t)$ = Velocity potential

$\phi(r, z)$ = Spatial velocity potential

r, θ, z = Radial, circumferential and axial coordinates

∇^2 = Laplace operator

ϕ = Scalar function

$W(r)$ = Mode shape of the plate

f_a = Natural frequency of the plate

f_l = Natural frequency of the plate in contact with the fluid

T_p^*, V_p = Reference kinetic energy and maximum potential energy of the circular plate

T_l^* = Reference kinetic energy of the fluid due to the motion of the plate

ω = Frequency in radians per second

β = Added virtual mass incremental factor (AVMI factor)

Γ = Non-dimensionalized added virtual mass incremental factor (NAVMI factor)

h_0 = Thickness of the center of the plate

μ = Taper parameter of the varying curve

n = Order of the mode number

N = Number of terms used in the interpolation function to approximate the mode shape

References

1. Kwak, M., *Hydroelastic vibration of circular plates.* Journal of Sound and Vibration, 1997. **201**(3): p. 293-303.
2. Jeong, K.-H., *Free vibration of two identical circular plates coupled with bounded fluid.* Journal of Sound and Vibration, 2003. **260**(4): p. 653-670.
3. Bathe, K.-J., H. Zhang, and S. Ji, *Finite element analysis of fluid flows fully coupled with structural interactions.* Computers & Structures, 1999. **72**(1): p. 1-16.
4. Heil, M., *An efficient solver for the fully coupled solution of large-displacement fluid–structure interaction problems.* Computer Methods in Applied Mechanics and Engineering, 2004. **193**(1): p. 1-23.

5. Kwak, M. and K. Kim, *Axisymmetric vibration of circular plates in contact with fluid.* Journal of Sound and Vibration, 1991. **146**(3): p. 381-389.
6. Amabili, M. and M. Kwak, *Free vibrations of circular plates coupled with liquids: revising the lamb problem.* Journal of Fluids and Structures, 1996. **10**(7): p. 743-761.
7. Amabili, M. and M. Kwak, *Vibration of circular plates on a free fluid surface: effect of surface waves.* Journal of Sound and Vibration, 1999. **226**(3): p. 407-424.
8. Liu, M.-F. and T.-P. Chang, *Axisymmetric vibration of a varying-thickness circular plate in contact with fluid.* Mechanics based design of structures and machines, 2004. **32**(1): p. 39-56.
9. Roszak, R., et al., *Fluid-structure interaction for large scale complex geometry and non-linear properties of structure.* Archives of Mechanics, 2009. **61**(1): p. 3-27.
10. van Opstal, T., et al., *A finite-element/boundary-element method for large-displacement fluid-structure interaction.* Computational Mechanics, 2012. **50**(6): p. 779-788.
11. Chen, J. and Q.S. Li, *Analysis of Flutter and Nonlinear Dynamics of a Composite Laminated Plate.* International Journal of Structural Stability and Dynamics, 2016. **16**(06): p. 1550019.
12. Shao, C., et al., *Flutter and Thermal Buckling Analysis for Composite Laminated Panel Embedded with Shape Memory Alloy Wires in Supersonic Flow.* International Journal of Aerospace Engineering, 2016. **2016**: p. 12.
13. Yao, G. and F.-M. Li, *Stability and vibration properties of a composite laminated plate subjected to subsonic compressible airflow.* Meccanica, 2016. **51**(10): p. 2277-2287.
14. Meirovitch, L., *Elements of vibration analysis.* 1975: McGraw-Hill.

Effects of Using Graphics and Animation Online Problem-Based Learning on Visualization Skills among Students

A Ariffin[1], M A Samsudin[2], A N Md. Zain[2], N Hamzah[1] and M E Ismail[1]
[1] Universiti Tun Hussein Onn Malaysia
[2] Universiti Sains Malaysia

E-mail: hasnida@uthm.edu.my

Abstract. The Engineering Drawing subject develops skills in geometry drawing becoming more professional. For the concept in Engineering Drawing, students need to have good visualization skills. Visualization is needed to help students get a start before translating into a drawing. So that, Problem Based Learning (PBL) using animation mode (PBL-A) and graphics mode (PBL-G) will be implemented in class. Problem-solving process is repeatedly able to help students interpret engineering drawings step work correctly and accurately. This study examined the effects of PBL-A online and PBL-G online on visualization skills of students in polytechnics. Sixty eight mechanical engineering students have been involved in this study. The visualization test adapted from Bennett, Seashore and Wesman was used in this study. Results showed significant differences in mean scores post-test of visualization skills among the students enrolled in PBL-G with the group of students who attended PBL-A online after effects of pre-test mean score is controlled. Therefore, the effects of animation modes have a positive impact on increasing students' visualization skills.

1. Introduction

Visualization is a very important skill in the subjects related to engineering (Leopold et al., 2001). In the industry, one of the components of graduate employability is able to generate new ideas in developing a quality product (Polytechnic Graduate Executive Report, 2008). Engineering Drawing is a major problem in most of the students because they do not understand the concept clearly and they are weak on visualization skills (Mohd Khairulanwar, 2010; Widad &Lee, 2004). Students have difficulty to imagine a 3D object into 2D (Nazamuddin, 2003). The failure is caused by a complex theory in certain topics and students depend on the lecture-answer method (Abdul Aziz, 2008; Yue, 2006; Tang, 2005). Mohd Salleh et al. (2011) said that in the Engineering Drawing subject, students had difficulty in understanding Isometric drawing topic. Approaches or methods like Problem Based Learning (PBL) can create great ideas and encourage students to build knowledge more meaningful (Joseph & Razmah, 2006). Engineering Drawing promotes culture of innovation and the ability to think visually in the product creation process from the beginning to the end (Muhyiddin, 2010; Zain, 2010).

Integration of online PBL can provide a platform to stimulate active and student-centered learning (Bourne, Harris & Mayadas, 2005). Online learning is a learning environment that uses the Internet network environment that helps students get the information, knowledge and skills in cyberspace (Fauziah, Hanafi & Guan, 2010). This method supports the traditional practice of face-to-face interaction in the classroom. This research shows PBL online learning can promote the characteristics of collaborative and self-directed learning (Brodie & Porter, 2008). This method is more flexible and suitable for a variety of learning styles.

One of the reforms in the PBL method in this study is online learning. The addition of PBL online is a synchronous discussion where students will discuss through a chat room and forum (Fauziah, Hanafi & Guan, 2003; Bernard & Rubalcava, 2000). This method is more efficient whereby mature students explore the links to web based sources to look for information (Bol & Garner, 2011). The findings from Megat Shahrul Anuar and Zahiri Aman (2004) showed that online learning has received encouraging response and is able to support students in their learning minds. The efficient transfer of knowledge is a very difficult challenge to educators. This means that, an educator had to adapt to the development and application of electronic teaching media.

Problem-Based Learning Model by Fogarty (1997) is applied in this study as a systematic problem-solving process and the problem will be seen repeatedly in every step. Figure 1 shows the eight main steps in the PBL model adapted from Fogarty's model.

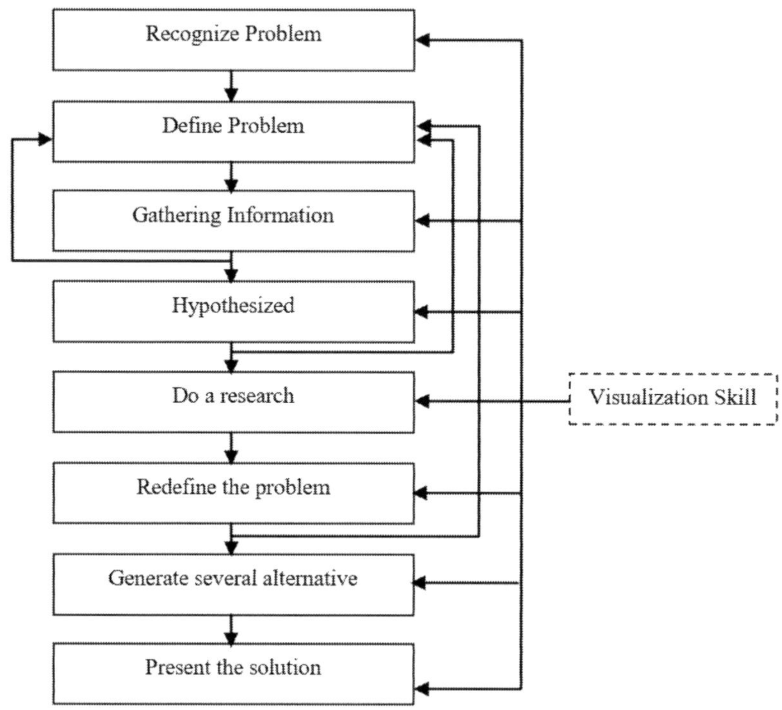

Figure 1: Problem-Based Learning Model

In the first step, the students will present their storytelling with a problem situation displayed in graphics mode and animation mode. The students will watch a short video and read the display on the e-engineering Drawing blog. The issue resembles real life and unstructured problem (Rio, 2008). Animation and graphic presentation help students to improve their visualization skills. The next step, students will define the problem based on existing information and experience. Various definitions of treatment will be given by the group. Definition of the problem will be changing throughout the process. In a step to collect information, students will use existing knowledge and experience. FILA table is used to facilitate in gathering information of students and their plan for the next solution. FILA offers four separate columns that need to be filled by students of a 'fact', an 'Idea', a 'learning issues' and a 'source needed'. This table is reviewed together by a group of students and lecturers.

Information collected by the students will be reviewed on an ongoing basis. Students will review the problem and predict the solution to build a hypothesis. The search begins to focus on the study hypothesis. In the next step, students will undertake a study to explore the information through books, websites and making opinions. Students will ask for help if they need from lecturers. Information and ideas are shared with the members of the group. Again, the students are asked to define the problem. A collection of information will change the definition of the problem in each group. Lecturers will monitor the learning process. In step seven, students are given the command to generate a suitable alternative to the solution of the problem. There are three categories of alternative proposed possible solutions, and solutions that can be a better solution. Next, the students will conduct group discussion and make the final decision. They will present arguments to support their findings.

The PBL problem situation may be presented online via storytelling and visualized through animation mode and graphics mode as articulated by Mayer (2001). Table 1 shows the three assumptions in his theory which are dual-channels of information processing, limited capacity and active-processing (Abd. Hadi, 2005).

Table 1: Mayer's Theoretical Assumptions

Assumptions	Explanation
Dual-channel assumption	The first assumption is the unproven but somewhat accepted theory that we process visual and verbal information differently and in two separate channels. Known as Paivio's Dual Coding Theory, it states that we process and internally represent visual information in a different way than verbal information.
Limited capacity assumption	The second assumption is that due to the capacity of working memory, we can only process a limited amount of information in each channel at one time.
Active-processing assumption	The final assumption is that in order to make something meaningful, people actively process information by paying attention and organizing and integrating the information.

Mayer's theoretical assumptions were developed based on Sweller's Cognitive Load Theory, Paivio's Dual Coding Theory and Baddeley's Working Memory Model (Fatimah, 2006). Based on the sensory-modality approach in Baddeley's Working Memory Model, instructional materials can be in the visual display (animation, graphics, text) or auditory (narration or sound). With reference to Paivio's Dual Coding Theory, Mayer (2001) divides the channel into two types of information presentation of verbal and visual. This theory proposes that verbal and visual materials are processed in different processing system of the auditory canal/ verbal and visual channels/ pictorial (Abd. Hadi, 2005). Figure 2 shows the Mayer's theory on the auditory / verbal canal and visual / pictorial channels.

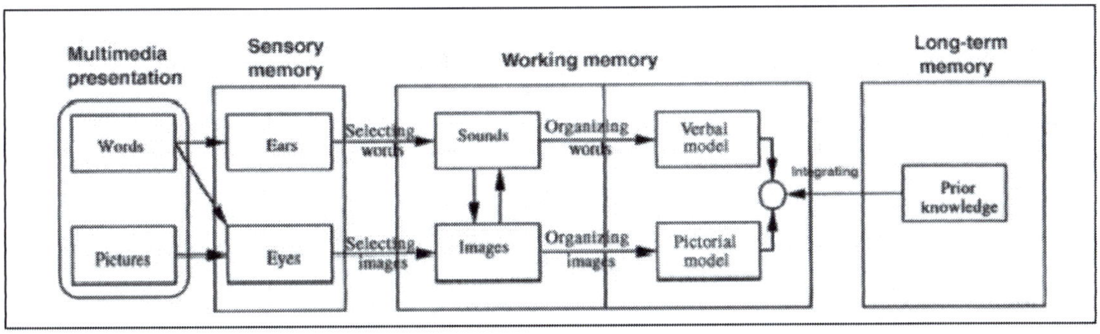

Figure 2: Cognitive theory of multimedia learning

Bertoline (1998) defines visualization as a person's ability to create, manipulate and interpret the images in the mind. Based on this definition, a person who is imagining a design, pattern or certain objects have never even seen physically is said to have some visualization capabilities. Similarly, a person who is able to imagine a shape, pattern and object to customize the environment and the time varies from place and time of original so have some abilities.

In engineering, a turning abilities, transform (Sorby et al., 2005), folding and stretching (Engineering Drawing, 2003), and deleting objects mentally (Field, 1999), is part of spatial ability. This capability is not easy, especially involving the thought of an object that has not been seen in front of the eye and something that has not yet physically exist. Psychological aspects of education, skills, visualization are understood as a natural ability in describing a person even without formal training or correlates with spatial ability (Sorby, 1999). In the other words, all human beings have this ability since they were born. However, the level of visualization skills may vary from person to person as with other skills.

1.1 PBL-A and PBL-G Based on Theory and Model

In this phase, researchers adapt Constructivist Theory, Cognitive theory of multimedia learning (2001) and Problem-Based Learning Model by Fogarty (1997) as the main essence in PBL-A and PBL-G online. PBL-A and PBL-G online consist of eight main steps: (1) identify the problem, (2) define the problem, (3) collect information, (4) creates a hypothesis, (5) do research, (6) redefine the problem, (7) generate some alternatives and (8) suggest solutions. Narrative design problems using animation mode and graphics mode were developed based on the Cognitive theory of multimedia learning (2001). Media content learning aims to produce narrative problems using text, graphics, audio, and video animation. The typeface used is a sans serif typeface, in Arial 12. The graph is loaded in JPEG format while the audio recordings in MP3 format. Video animation learning materials were developed in FLV format, while PBL teaching materials were developed using Blendspace website, Wiki, Blog, Mind Mapping, Poll, Dropbox, Glogster and pirate pad. In this study, design and website development of e-Engineering Drawing is based on Constructivist Theory and Problem-Based Learning Model by Fogarty (1997). Metacognitive process will start if new information can be understood by students. Therefore, researchers should take into account the learning process that is more efficient and can enhance students' understanding of the issues highlighted in the PBL-A and PBL-G. Integration of animation and graphic storytelling problem with the situation is a major step to help the students' develops metacognitive processes. Next, students can build knowledge, new ideas and concepts in learning in Engineering Drawing. Problem-Based Learning Model by Fogarty (1997) was adapted in the website e-Engineering Drawing.

Problem-Based Learning Model by Fogarty (1997) was adapted to design the website to facilitate the students' visualization through the process of PBL-A and PBL-G. Cognitive theory of multimedia learning (2001) was used to clarify the situation in storytelling problem for PBL online using graphics mode and animation mode. This theory provides a theoretical support for the use of visuals, whether dynamic or static in a learning offering (Paivio, 1986). Mayer's theory affects the information presented visually or auditory. Mayer (2001) divides the channel into two types of information presentation, verbal and visual. This theory proposes verbal and visual materials processed in different processing systems of the auditory canal / channel verbal and visual / pictorial (Abd. Hadi, 2005). Based on this theory, researchers design problems connecting narrative information in verbal and visual,

i. PBL-G: Graphic with text
ii. PBL-A: Animation with audio

Strategies and approaches in verbal and visual information presentation are also important to ensure the effective application of information in long-term memory. Animation and graphic designs use Adobe Photoshop CS2, Illustrator CS and Macromedia Flash 8. Some of Cognitive theory of multimedia learning (2001) principles serve as a guide in designing PBL-A and PBL-G. Here are the principles underlying the Cognitive theory of multimedia learning (2001),

i. Multimedia principles
Students can learn better through the use of text and images of text only. Therefore, researchers combine the use of text along with pictures to enhance the effects of better learning.

ii. The principle of proximity space
Students can learn better if the text associated with the image is placed closer together than apart. In PBL-G, travelled graphics along with the related text to enable students to understand the situation of the problems raised.

iii. The Principles of time
Students can learn better if the text with images is displayed simultaneously compared to turns.

iv. The principle of coherence
Student learning will be disrupted if the text and pictures are interesting but not relevant to the content of the lesson. Similarly for a multimedia presentation with audio interrupted if unrelated. A better way is to get rid of the text and audio researchers which are irrelevant to enhance the learning effect.

v. Modality principle

Students can learn better if there is an animation with audio along with the text. PBL-A is based on the use of animated elements and associated audio.

vi. The Principles redundancies

Students can learn better if there is an animation with audio of along with text. This increases the cognitive load when combined together. The researcher did not incorporate animation, audio and text to avoid confusion when the intervention was implemented.

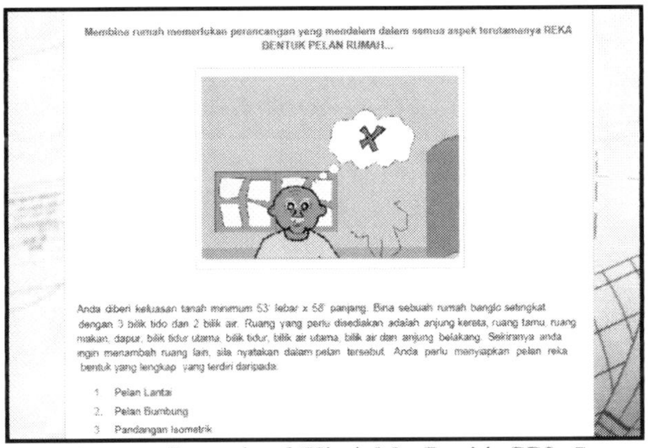

Figure 3: Example of 'Kisah Mat Bon' in PBL-G

Figure 4: Example of 'Kisah Mat Bon' in PBL-A

2. Methodology

The objective of this study is to compare the effects of PBL-A online and PBL-G online on visualization skills of students in polytechnics. In this study, the selected design is quasi-experimental study in the form of pre-test and post-test of the two treatment groups. PBL group online through trigger in graphic mode is called PBL-G and the PBL online through problems in the mode of animation storytelling is called PBL-A. Further, these groups took a post-test II to identify the retention of new learning methods (Table 2).

Table 2: Group Experiment Design

Group	Intervention	
PBL-G	O_1X_1	O_2 .... O_3
PBL-A	O_1 X_2	O_2 .... O_3

Which is:
x_1 = PBL-G
x_2= PBL-A
O_1 = Pre Test
O_2 = Post Test I
O_3= Post Test II

Purposive sampling is used in this study. This refers to a sampling procedure in which a group of subjects has certain characteristics as respondents in the study (Piaw, 2006). Sixty-eight Mechanical Engineering students from polytechnic were selected. This study only focused on the Isometric Drawing topic which covering a three month period. The instrument used is visualization test adapted from Bennett, Seashore and Wesman. The reliability coefficient was 0.91 for the entire item. Training in aptitude tests was selected because each item is built on the concepts of Engineering Drawing. There were 60 items and each one has a different figure. Each item in the test covers isometric, spreadsheets, reflection, sectional and combinations. Students were given 25 minutes to answer all the questions. The test is used to determine the ability of students to fold 2D objects to 3D objects mentally (Mohd Salleh, 2011).

3. Result and Discussion

The findings of the analysis of the PBL-G found that post-test mean score (M =36.69, SD =8483) were relatively higher than the pre-test mean score for graphic visualization (M =24.38, SD =8419). Please refer to Table 3.

Table 3: Description of the mean, standard deviation, minimum and maximum

	Pre-test	Post-test
N	32	32
Mean	24.38	36.69
Standard deviation	8.419	8.483
Minimum	15	17
Maximum	52	49

Table 4 show the post-test mean score of the PBL-A (M = 42.36, SD = 7453) were relatively higher than the pre-test mean score on animation visualization (M = 26.67, SD = 6476).

Table 4: Description of the mean, standard deviation, minimum and maximum

	Pre-test	Post-test
N	36	36
Mean	26.67	42.36
Standard deviation	6.476	7.453
Minimum	14	26
Maximum	43	57

The analysis showed that there was an increase on visualization skills for the groups which used animation mode PBL-A compared to graphics mode in PBL-G. This finding is consistent with studies of Norhayati, Shaferul and Mohd Fauzi (2013) who found that the use of animation helps students apply the information provided. PBL-A is designed to emphasize PBL Fogarty's Model (1997) in repeated trigger. Every step of the PBL may have to be reconsidered. The situation of storytelling problems is displayed in animation mode. As a result, students will be able to create and manipulate information of storytelling in animation problems. This helps improve their visualization skills compared to the use of graphics. Based on the theory of Mayer, graphics gives less impact as compared to the use of animation. Indirectly, PBL also promotes learning through problem solving through the application of information effectively (Harris & Briscoe-Andrews, 2008).

Arumugam (2006) found that this approach can improve students learning through animation and increase their curiosity. PBL scenario is focused to use students' curiosity so that problems can be solved (Hmelo-Silver, 2004). Through PBL-A, visualization abilities of students can be developed through computer animation, internet, website and online learning (Abdul Hadi, 2005). It also coincides with the views of Jamaluddin and Zaidatun (2003) who found that animation helps accelerate the process of understanding and maintaining information within a long time in the student's memory.

Through a one-way ANCOVA analysis, the impact of PBL is applied to the values of mean visualization which skills test can be determined. The results of this analysis are shown in Table 5. One-way ANCOVA analysis of the data showed that the F=8348. It means squared error=536 163 and $p<0.05$. This shows that there are significant differences in post-test which means score visualization skills between students who attended PBL-G online group was pursuing PBL-A online and after effects of pre-test mean score of controlled visualization skills.

Table 5: One-way ANCOVA to test mean score visualization skill

Source	Type III Sum of Squares	df	Mean Square	F	Sig.
Corrected Model	545.612[a]	2	272.806	4.247	0.018
Intercept	8209.057	1	8209.057	127.809	0.000
Pre_visualization	0.278	1	0.278	0.004	0.948
Group	536.163	1	536.163	8.348	0.005
Error	4174.903	65	64.229		
Total	111847.000	68			
Corrected Total	4720.515	67			

Visualization is a technique used in creating the image or animation that communicates with a message to be conveyed (Knight, 2000). This finding is also supported by Ahmad Zamzuri (2012). In his study, it is found that animation can cause teaching and learning process to be more efficient rather than the use of graphics. He added that animated presentation techniques can move the channels in verbal and visual memory and is an effective strategy for acquiring, processing, store and recall information. In this study, the effects of animation in the Mat Bond story could explain the problems that arise when building a house is not planned well. Mat Bond's movements and his expressions can describe Mat Bond's character in real situations. Problems arise when the size is not in accordance with the standards and the plans are not in accordance with the requirements. Problem situation helps to improve students' skills in visualization. In addition, the use of animation augmented with audio assistance can improve students' learning effects. This is in line with the principle of Mayer's theory where animation and sound presentation give a better impression than the graphical presentation along with text (Mayer, 2001). The relationship between input visual or verbal input or a combination of both gives the opportunity to build understanding through visual depiction income in the minds of students.

This finding is also consistent with studies of Rosiadi et. al. (2009) who designs teaching aids by using animation and sound element's background that explains the process of learning. Lindstrom (1994) argued that the animations bring users closer to material information. The results of the study clearly show that their visualization skills will also increase. Students will also generate new ideas through their minds with the impression generated in the form of graphs or drawings. Infused storytelling animation problems are repeatedly used at every step of Fogarty's PBL Model. This is to reinforce information to be conveyed. This helps students to structure the knowledge, and further understand the teaching content presented more clearly. PBL approach for online students gives access information quickly and facilitates collaboration among online communities (She &Emy, 2007). This strategy can enhance students' knowledge and skills to contribute to the improvement of visualization. Therefore, students who attend PBL-A can build a very meaningful experience in learning and help train the skills to become a better visualization.

Through a one-way ANCOVA analysis, the impact of PBL is applied to the values of the mean post-test II visualization skills can be determined. The results of this analysis are shown in Table 6. One-way ANCOVA analysis of the data showed that the value of F=11.969, mean squared error=376 862 and $p<0.05$.

Table 6: One-way ANCOVA to test mean score visualization skill

Source	Type III Sum of Squares	df	Mean Square	F	Sig.
Corrected Model	378.228[a]	2	189.114	6.006	0.004
Intercept	12325.831	1	12325.831	391.461	0.000
Pre_visualization	42.541	1	42.541	1.351	0.249
Group	376.862	1	376.862	11.969	0.001
Error	2046.640	65	31.487		
Total	169339.000	68			
Corrected Total	2424.868	67			

Integration of PBL-A in the process of teaching and learning in polytechnics can tackle weaknesses in students' visualization. The findings are also consistent with studies of Fong, Kong and Fauzy (2001) which found that the use of text with animated shows a positive trend compared to the use of static text with graphics. Based on Mayer's theoretical principle, PBL-A is designed for the situation of storytelling problems that are displayed visually or auditory (speech) along with the sound. It allows something rather difficult to explain with words or static images more easily and effectively (Jamalludin & Zaidatun, 2003). Mayer (2001) found that the first information which will be directed is then followed by animated images, text and so on. Animation also affects how students use information clearly in software (She & Chen, 2009). Therefore, it is clear that the potential in animation stimulate students' focus on such information. Wang (2012) states that integration problems in a realistic animation contribute in a more meaningful experience than the group of students who use the help text only. Integration of animation modes in PBL-A is more dynamic and interactive features to help students sharpen their problem-solving process better.

Mohd Salleh (2011) found that students like to sketch drawing and moving images by using animation. Teaching aids which are built using animated elements also have a positive impact in the process of teaching and learning and visualization skills to maintain them in the long run. Online discussion space between team members and question-answer session in forums helps them to expand existing knowledge and improve their visualization skills. They realize that with PBL, they need to have a better understanding of the problem and how to solve it (Colliver, 2000). Mohd Khairulanwar (2010) added that plane geometry and painting based on a realistic visual contributes to the observation of implicit forms. For example, in the construction of isometric, students need to associate the concept of perceived experiences and draw a visual representation of the real (Abu Bakar, 1991). Observation of a problem needs to be done to get a more in-depth input and it is considered important. Indirectly, knowledge and problem-solving skills of students will help increase the visualization.

4. Conclusion

The study found that PBL-A line had a positive effect compared to PBL-G online. In conclusion, the individual is able to achieve good visual cognitive level. They master the knowledge, ideas and new approaches easily. Students will continue to build new knowledge from the information obtained with existing knowledge of students (Rio Sumarni, 2006). Curiosity among students has also increased and they will achieve a higher level of thinking as required. Students are no longer passive, but active in solving problems. A PBL helps students master skills, especially related to visual or shadow. They will realize that in this method, they need to understand the problems better by using animation and figures as a solution to the problem. (Colliver, 2000). Problems created to allow students to explore their own solutions while improving visualization skills and their academic achievement. Therefore, engineering students are able to achieve the desired educational goals.

5. Acknowledgment
This research was partially supported by University Tun Hussein Onn Malaysia, Johor using Short Term Grant (U646). We thank our colleagues from University Sains Malaysia who provided insight and expertise that greatly assisted the research.

6. References

Abdul Hadi Mat Dawi (2005). *Kesan animasi 3D terhadap pembelajaran Lukisan Ortografik di kalangan pelajar sekolah menengah.* Tesis PhD. Universiti Sains Malaysia.

Ahmad Zamzuri Mohamad Ali (2012). Lima Prinsip Persembahan Animasi dalam Perisian Pendidikan. *Jurnal Teknologi Pendidikan Malaysia.* 1(2), 15-22.

Arumugam, R. (2006). *Kesan Peta Animasi Dan Interaktif Dalam Pengajaran Dan Pembelajaran Geografi.* Tesis PhD. Universiti Pendidikan Sultan Idris.

Bernard, J. & Rubalcava, A. (2000). *Technical Graphics Communication* (3 ed.). New York: McGraw-Hill.

Bol, L., Joanna, K. & Garner (2011). Challenges in Supporting Self-Regulation in Distance Education Environments. *Journal Computer High Education.* 23. 104-123.

Bourne, J., Harris, D. & Mayadas, F. (2005). "Online Engineering Education: Learning Anywhere, Anytime." *Journal of Engineering Education.* 94(1), 131-146.

Brodie, L.M, & Porter, M (2008). Engaging distance and on-campus students in problem based learning. *European Journal of Engineering Education.* 83(3), 135-140.

Colliver, J. A. (2000). Effectiveness of problem-based learning curricula: research and theory. *Journal of Education.* 75 (3).

Fong, S.F., Kong, W. N. & Fauzy, W.M. (2001). *Effects of computer graphic presentation among students of different anxiety levels.* Proceeding 14th Educational Technology Convention pp 44-60 at Hotel Goldcourse, Kelang, Malaysia 11 – 14 Sept 2001.

Harris, A. & Briscoe-Andrews, S. (2008). Development of a problem-based learning elective in "green engineering". *Education for Chemical Engineers.* 3(1), 84-91.

Hmelo-Silver, C. E. (2004). Problem-based Learning: What and How Do Student Learn?. *Educational Psychology Review.* 16 (3). 235-266.

Jamalludin Harun & Zaidatun Tasir (2003). *Multimedia Dalam Pendidikan.* Bentong: PTS Publications & Distributors Sdn. Bhd.

Jabatan Pengajian Politeknik Malaysia (2011). *Silibus Mata Pelajaran Lukisan Kejuruteraan.* Putrajaya: Jabatan Politeknik Malaysia.

Kurikulum Bersepadu Sekolah Menengah (1993). *Sukatan Pelajaran Kurikulum Bersepadu Lukisan Kejuruteraan.* Kuala Lumpur: Pusat Perkembangan Kurikulum.

Leopold, D.A., O'Toole, A.J., Vetter, T., and Blanz, V. (2001). Proto-The orientation-selective adaptation control was carried out on type-referenced shape encoding revealed by high-level after effects. seven naive observers using the same setup as in experiment 3. *Nat. Neurosci.* 89–94 (4).

Mayer, R. E. (2001). *Multimedia learning.* New York: Cambridge University Press.

Megat Aman Zahiri dan Shahrul Anuar (2004) *Animasi 3D Digital: Alatan Kognitlf bagi Meningkatkan Prestasi Visualisasi Mental dalam Pendidikan untuk Pembangunan Lestari.* Seminar Pendidikan: Pendidikan untuk Pembangunan Lestari, Hotel Shangri-La Pulau Pinang.

Mohd Khairulanwar Abdul Khalil (2010). *Modul Pembelajaran Kendiri Berbantukan Komputer bagi Lukisan Isometri.* Tesis Sarjana Muda. Universiti Teknologi Malaysia.

Mohd Salleh Tahar (2011). Keberkesanan Penggunaan CD Interaktif Lukisan Isometrik Terhadap Kemahiran Visualisasi Pelajar. *Jurnal Pendidikan Teknik dan Vokasional.* 104-115 (3).

Muhyiddin Mohd Yassin (2010). *Tranformasi Politeknik.* Putrajaya International Convention Centre. 25 Februari 2010.

Nazamuddin Ahmad Razali (2003). *Penghasilan Modul Pembelajaran Lukisan Kejuruteraan Berbantukan Komputer.* Tesis Sarjana. Universiti Tun Hussein Onn Malaysia. Tidak diterbitkan.

Norhayati Che Hat, Shaferul Hafes Saari & Mohd Fauzi Abdul Hamid (2013). Persepsi Pelajar Terhadap Penggunaan Animasi dalam Pembelajaran Bahasa Arab. *Jurnal Teknologi.* 25-29.

Piaw, C. Y. (2006). *Statistik Penyelidikan Lanjutan: Ujian Univariat dan Multivariat.* Kuala Lumpur: McGraw-Hill Sdn. Bhd.

Polytechnic Graduate Executive Report (2008). *Jabatan Pengajian Politeknik dan Kolej Komuniti.* Kuala Lumpur: Kementerian Pengajian Tinggi Malaysia.

Rio Sumarni Sharifuddin (2009). *Kerangka Bahan Instruksi Inovatif Berasaskan Penyelidikan.* Skudai:Nasmax.

Rosaidi Shukri, Faridzul Fahmi, Noorashikin Md Nor & Huzaima Amin (2009). *Kajian Kes Aplikasi Model Reka Bentuk Instruksional dalam Proses Pengajaran dan Pembelajaran.* Diakses pada 29 Februari 2009 dari *http://www.scribd.com/doc/3682723/Kajian-Kes-Aplikasi-Model-Rekabentuk-Instruksional-dalam-PP-Koswer-Pendidikan*

She, Y. & Chen, J. (2011). Effect of problem-based, self directed undergraduate education on life-long learning. *Canadian Medical Education Journal.* 148, 999–976.

Sorby, S. A., (2003). *Developing 3-D Spatial Visualization Skills for Engineering Students.* American Society for Engineering Education Annual Conference & Exposition, Portland, Oregon.

Tang, H. E. (2005). *Kesan Sistem Pembelajaran Berkomputer Matematik Berasaskan Personaliti Serta Kemahiran Berfikir Secara Kritis dan Kreatif.* Tesis PhD. Universiti Teknologi Malaysia. Tidak diterbitkan.

Wang, H. (2012). The comparative efficacy of 2D- versus 3D-based media design for influencing spatial visualization skills. *Computers in Human Behavior.* 1943–1957.

Widad Othman & Hatta Ismail (2001). Aplikasi Konsep Dan Kaedah Lukisan Kejuruteraan Dalam Matematik Oleh Guru-Guru Matematik Di Sekolah Menengah Teknik Seluruh Negeri Kelantan. *Jurnal Teknologi.* 35. 1-10.

Yue, K. N. (2006). *Jump Start Authentic Problem-based Learning.* Prentice Hall: Pearson Publication.

Yusup Hashim & Razmah Man (2006). *Teknologi Instruksional: Teori dan Aplikasi.* Shah Alam: Fajar Bakti.

Zain Retas (2010). *Impak Pertandingan Inovasi Projek Pelajar Ke Arah Peningkatan Tahap Keyakinan Diri dan Kesediaan Pelajar untuk Menempuhi Alam Pekerjaan.* Di akses pada 15 November 2011 dari http://www.pmm.edu.my/

IOP Conf. Series: Materials Science and Engineering **203** (2017) 012023 doi:10.1088/1757-899X/203/1/012023

The use of animation video in teaching to enhance the imagination and visualization of student in engineering drawing

Ismail, M.E., Irwan Mahazir, I., Othman, H., Amiruddin, M.H., & Ariffin, A.
Faculty of Technical and Vocational Education
Universiti Tun Hussein Onn Malaysia

E-mail: erfy@uthm.edu.my

Abstract. The rapid development of information technology today has given a new breath toward usage of computer in education. One of the increasingly popular nowadays is a multimedia technology that merges a variety of media such as text, graphics, animation, video and audio controlled by a computer. With this technology, a wide range of multimedia element can be developed to improve the quality of education. For that reason, this study aims to investigate the use of multimedia element based on animated video that was developed for Engineering Drawing subject according to the syllabus of Vocational College of Malaysia. The design for this study was a survey method using a quantitative approach and involved 30 respondents from Industrial Machining students. The instruments used in study is questionnaire with correlation coefficient value (0.83), calculated on Alpha-Cronbach. Data was collected and analyzed descriptive analyzed using SPSS. The study found that multimedia element for animation video was use significant have capable to increase imagination and visualization of student. The implications of this study provide information of use of multimedia element will student effect imagination and visualization. In general, these findings contribute to the formation of multimedia element of materials appropriate to enhance the quality of learning material for engineering drawing.

1. Introduction

The process of teaching and learning is an important element in ensuring that the students can follow the content of their subject. Teaching is an activity related to the diffusion of knowledge covering the planning, management, delivery, supervision and evaluation. The virtual learning has become a new method in the process of teaching and learning in order to facilitate the delivery of content. Virtual learning is learning by using electronic media and Internet usage. E-Learning is an acronym of Electronic Learning, it is a new way of teaching and learning using electronic media, especially the Internet as a learning system. Besides, electronic multimedia can be an effective medium of communication and positive for its ability to simultaneously combine different colors and styles of text, audio, video and animation to be displayed on the screen [1].

However, this situation can be delivered through other means such as the use of teaching aids. The appropriate teaching aids usage in the process of learning make that students focus and understand well what is presented by the teacher [2]. With the relevant teaching aids, it can help teachers explain the concept more clearly than with a verbal description such as the use of video animation. Directly, this creativity can help teachers make teaching and learning easily to deliver the content to students in certain subjects such as engineering drawing subject more on practical teaching or known as hands-on.

2. Problem Background

Academic excellence and skills of students technical and vocational often associated with teaching methods by teachers and also students learning styles itself. Therefore, the core of excellence students may be located on the use of teaching aid by teachers. In line with advances in technology, internet usage priority to students currently and most students find information to supplement knowledge via the internet but information about teaching and learning related technical and vocational skills are still not enough.

In order to understand the content of practical lessons, students are required to perform tasks itself to the skills. Learning based on teacher impact that students only receive information presented by teachers and less in realizing education content on the subject is learned [3]. Thus, to get good results, teachers need to be clever and creative in producing teaching aids like animated video production to help teachers deliver lessons with more content and be understood by students.

Most of secondary schools lack variety of infrastructure such as computer network to facilitate more subjects which is taught with the use of multimedia technology and teaching through a web site [4]. Some importance of the use of teaching aids based e-learning identified to increase the ability competency of students and be able to perform the assessment of teaching and learning approach that is used by the teacher. In addition, it can help the teacher in explaining the content of the lesson and can reduce boredom and less focus in teaching and learning sessions [5].

According to [6], teaching traditionally involves the use of text books and teacher centered. [7] felt that the traditional teaching and learning methods is 'chalk and talk' will cause the students facing difficulty to construct knowledge independently. Therefore, this strategy must be improved in order to create an active teaching and learning where students also thought to solve the problem. The requirements of ICT in teaching and learning for his role in improving the quality of presentation in the teaching and learning as well as help strengthen students understanding of a concept learned [8]. Therefore, teachers must prepare themselves with various computer skills as one of the key elements to assist and facilitate the process of teaching and learning in the classroom [9].

[10] view that teaching methods using wireless devices enabling learning to occur at any time, in particular in technical and vocational education. According to [11], the more senses used increasingly effect the learning. The view that the different senses produce different responses [11]. Each of the senses complement deficiencies and strengthen other senses.

Indeed, teaching aids plays an important role in the teaching and learning process of teachers including technical teachers. It is the responsibility of teachers to use them with best possible for help understanding his students about concepts and knowledge lessons to be served.

3. Problem Statement

The problem that exists is the use of static media such as text coupled with the difficulty of the content of engineering drawing will improve students cognitive load [12]. In view of the limited working memory capacity, students won't be able to focus on the content of education and the solution of a difficult assignment work simultaneously. Therefore, problems arise students understand the information presentation problem [13]. Student understanding and learning is also closely related to the teaching methods used by teachers that makes the matter as an issue that should be settled. [9] found that students often have difficulties in solving engineering drawing due to weak in visualization. [14], states that technological innovation in the classroom will hopefully be able to bring positive changes in students' academic achievement. According to [6], teaching traditionally involves the use of textbooks and teaching methods in class only will result in students easily feel bored and has dampened the interest of students to learn. Therefore, this study maneuverability testing video animation to assist students in terms of understanding in enhancing the skills of visualization and imagination.

4. Methodology

The research design for this study was a survey method using a quantitative approach. The advantage of a survey is the data can be collected directly from respondents and the results can be generalized to the population. The samples consisted of students from the Vocational College at Setapak . A total of 30 students from Industrial Machining courses were selected randomly as respondents. A survey questionnaire developed by the researchers was utilized to collect data in order to investigate used the use of multimedia element for animated video in six elements, namely text, audio, animation, graphics, visualization and imagination. Then, the data collected were analyzed by using mean and standard deviation. A five point Likert scale was used (5= strongly agree – 1= strongly disagree) and was completed by the participants. A correlation coefficient value (0.83), calculated on Alpha-Cronbach, was used to test the internal reliability of the questionnaire and it showed an acceptable reliability value. Descriptive statistics were used to determine the use of multimedia element and describe students' skills of visualization and imagination toward engineering drawing.

5. Data Analysis

In specifically, respondent of (N=30) students strongly agree that the use of multimedia element very important for video animation that can be show questionnaire item text (mean=3.67, standard deviation=0.47), audio (mean=3.72, standard deviation=0.45), animation (mean=3.63, standard deviation=0.49) and graphic (mean=3.68, standard deviation=0.46). Furthermore, there are several indicators in the survey (Table 1) related to describe students' skills of visualization and imagination. The majority of respondents in this study confirmed the use of multimedia element will effect students' skills of visualization (mean=3.76, standard deviation=0.44) and imagination (mean=3.70, standard deviation= 0.46). This shows the decision of the respondent involved giving a response to the questionnaire agreed that the use of video animation able to give more understanding to the students. According to [15], video-based learning and additional shows the effectiveness of the training system the same learning to face to face between teachers and students. The use of technology can help teachers improve their teaching effectiveness, especially in making a visual concept, and motivation [16].

To assess the development of video animation in helping students to master and improve the ability of imagine the objects and their visual object on the drawing. The findings show the results of the respondents involved in providing answers and found that the mean value of the items is at an all-time high. The majority of respondents strongly agree on the items given and this shows the video animation is able to assist respondents in enhancing the imagination and visualization. [16] has stated that the video could help students imagine object. According to [17] the use of video can increase motivation and student learning outcomes. With the help of video learning can increase student learning activities and outcomes in learning [18].

The findings about the content in the video animation shows the results of the respondents is at high level. The majority of respondents indicated that they were very amenable to items given and showed video animations can help respondents in the process of learning. This indicates that respondents agree to the content in the video animation includes the contents of the subjects of engineering drawing and taking them into existing knowledge. Overall, video animation has fulfilled the syllabus subjects engineering drawing and was able to achieve the objectives of the subjects of engineering drawing and testing the existing knowledge of the respondent.

Table 1: Elements of engineering drawing animated video that enhancing the skills of visualization and imagination

Element	Mean	Std. Deviation
Text	3.67	0.47
Audio	3.72	0.45
Animation	3.63	0.49
Graphic	3.68	0.46
Visualization	3.76	0.44
Imagination	3.70	0.46

6. Discussion

Video animation for the teaching which is developed must have attractive presentation design. The use of pictures, graphs, diagrams, videos and demonstrations is suitable for the student category of visual [19]. The use of multimedia like video animation can increase the motivation of students to master learning outcomes and enhance the understanding [20]. With the use of this animation video, students are able to master the learning outcomes easier and more effective. [21] noted that good animation enables students to learn something quickly, retain students longer memory, present a clearer and accelerate students understanding. This animated video design is attractive and has obtained the consent of the respondent on the basis of the results provided through the questionnaire which is at high level. The respondent has shown a high level of agreement on the video animation able to give a better understanding. Overall, the design of a presentation video animation was produced this is appropriate with an average mean score is at a high level [22]. This shows the design presentation video animation already built this is appropriate to be used as teaching aids for subjects and Engineering Drawing indirectly has answered the first question of the study.

To ensure that the video was developed capable of helping in enhancing the imagination and visualization students, researchers have conducted an assessment of video animation by using two test which is tests alpha and beta testing [23]. After the improvement of alpha test, video animation next assessment is beta testing. The beta test is carried out on a sample of respondents. The result of the beta test was found that video animation is at high level. The findings of this beta testing show that this animation video appropriate and able to help students solve their imagination and visualization problem. Some improvements to the video animation done for usability of students after the beta test. Multimedia element that have an interactive visual graphics, text, music, video and animation, able to provide information of use of multimedia element will student effect imagination and visualization to who multimedia element in teaching methods [24]. According to [25], the integration of multimedia elements to give a session of teaching and learning more interesting, creative, innovative and provide learning experience effectively to students. Findings from the development of video animation shows respondents strongly agree with video animation which video this animation is able to assist students in improving their imagination and visualization. According to [26], the use of video animation can improve their imagination and visualization.

Engineering Drawing video animation evaluated the level of maneuverability to help students in the learning process. The use of technology in teaching and learning processes is undeniable because its potential in improving the academic achievement of students [27]. Through the analysis of the study showed an animated video produced can function properly and are at a high level. It is shown that multimedia is able to create a learning environment that is more attractive than conventional methods just based on text books [20]. Analysis has shown that video animations are able to assist students in improving the understanding during the learning process. The use of video animation is very different from the traditional learning. This is because the advantage in allowing a repeat topic that is not understood by the students can be carried out more easily with the use of video animation. According to [28], the use of animation is interesting and clearly give understanding to students in a very short time and can be repeated show for the purpose of strengthening the students understanding. Overall, it was found that this animation video is able to increase the understanding of the students in the learning process. In addition, the use of supplementary teaching materials by teachers during the process of learning and teaching will help improve the quality of cognitive and affective student [29].

7. Conclusion

Overall, the results of analysis found video animation developed meet the objective of the study which shows video design animation can increase student understanding. Development of video animation can also enhance the student's imagination and visualization implies structural work piece or products to be drawn. This animated video able to assist students in understanding teacher learning with the help of text, graphics, animation, audio and video that are able to increase the motivation of students and make learning process more interesting. Moreover, the selection of the media accurate and appropriate by the teacher are very closely related to the effectiveness of a process of teaching and learning. User friendly features found on the video animation that developed this can facilitate students to operate because the purpose of the development of video animation is to help students to increase imagination and visualization. Beside reduce the student's cognitive load the study contribute to the formation of multimedia element of materials appropriate to exchange quality of learning material. The Era technology nowadays will make students more savvy technology and makes it easier for students to use technology at school or at home. This situation caused the students prefer to use technology than text-module only can provide alternative learning material for engineering drawing course.

References

[1] Tasir, Z., & Harun, J. (2003). Asas Reka bentuk Aplikasi Multimedia Pendidikan. *PTS Publication*, 1–10.
[2] Mazur, E. (2013). Peer instruction.
[3] In'am, A., Saad, N. S., & Ghani, S. A. (2012). Pembangunan Model Pengajaran Dan Pembelajaran Matematik Berasaskan Metakognitif. *Jurnal Pendidikan Sains & Matematik Malayisa*, *2*(1), 23–37.
[4] Bakar, Z. A. (2008). Kemahiran ICT di kalangan guru pelatih IPTA Malaysia. Arah Pendidikan Sdn Bhd.
[5] Jamian, N. F. M. (2012). Kecenderungan gaya pembelajaran Model Dunn dan Dunn dan Model Honey dan Mumford di kalangan Pelajar Ijazah Sarjana Muda FPTP di UTHM (Doctoral dissertation, Universiti Tun Hussein Onn Malaysia).
[6] Ishak, A., Kasa, Z., Selamat, M. H., & Samah, B. A. (2009). Perbandingan Pengajaran Berasaskan Multimedia dan Tradisional ke Atas Pencapaian Matematik dan Sikap Matematik di Kalangan Pelajar Berisiko. *Jurnal Teknologi Maklumat Dan Multimedia*, *5*, 79–89.
[7] Purwanta, H. (2010). Penggunaan Penilaian Berbasis Kelas Untuk Meningkatkan Kualitas Proses Dan Hasil Belajar Siswa Kelas IV SD Negeri Samirono Yogyakarta Pada Mata Pelajaran IPS. *Jurnal Penelitian Vol*, *13*(2).
[8] Rusmini Ku Ahmad. (2003). Integrasi teknologi maklumat dan komunikasi dalam pengajaran dan pembelajaran. Jurnal Pengurusan dan Pendidikan, 13, 82–91.

[9] Yasin, R. M., Mustapha, R., Minghat, A. D., Jusoff, K., Ishar, A., & Shamsudin, S. (2012). Pelaksanaan Pengajaran Dan Pembelajaran Mata Pelajaran Lukisan Kejuruteraan Di Sekolah Menengah. *Asia Pasific Journal of Educators and Education*, *27*(1), 23–35.

[10] Abd Rahman, R., & Mohd Hashim, M. H. (2011). M-Pembelajaran Dalam Pendidikan Teknik Dan Vokasional (PTV) Di Malaysia. *Persidangan Kebangsaan Penyelidikan Dan Inovasi Dalam Pendidikan Dan Latihan Teknik Vokasional*, 1–11.

[11] Azman, M. N. A., Azli, N. A., Mustapha, R., Balakrishnan, B., & Mohd Isa, N. K. (2014). Penggunaan Alat Bantu Mengajar ke Atas Guru Pelatih Bagi Topik Kerja Kayu, Paip dan Logam. *Sains Humanika*, *1*(2000), 77–85.

[12] Baggott la Velle, L., Wishart, J., McFarlane, A., Brawn, R., & John, P. (2007). Teaching and learning with ICT within the subject culture of secondary school science. *Research in Science & Technological Education*, *25*(3), 339-349.

[13] Yap, C. S. (2012). Goal Orientation of Engineering Students and the Learning Strategies That They Employ. *3rd International PBL Symposium*. Republic Polytechnic, 7-9 March 2012.

[14] Gani, A. W. I., Siarap, K., & Mustafa, H. (2006). Penggunaan Komputer Dalam Pengajaran-Pembelajaran Dalam Kalangan Guru Sekolah Menengah: Satu Kajian Kes Di Pulau Pinang. *Kajian Malaysia*, *XXIV*(1), 203–225.

[15] Zhang, D., Zhou, L., Briggs, R. O., & Nunamaker, J. F. (2006). Instructional video in e-learning: Assessing the impact of interactive video on learning effectiveness. *Information and Management*, *43*(1), 15–27.

[16] Cheung, A., Slavin, R. E., Kim, E., & Lake, C. (2015). Effective Secondary Science Approaches : A Best-Evidence Synthesis, (June).

[17] Cahyani, R., & Karyanto, P. (2016). Penggunaan Media Video Untuk Meningkatkan Motivasi Dan Hasil Belajar Materi Biosfer Pada Siswa Kelas XI IPS, *2*(2).

[18] Setyosari, P., & Malang, P. D. P. (2016). Penerapan Pendekatan Savi Berbantuan Video Pembelajaran Untuk Meningkatkan Siswa Kelas Iv Sdn I Sanan, 1235–1241.

[19] Abdul Razak, R., & Abdul Rahman, M. (2013). Pembinaan Media Pengajaran Berasaskan Multimedia Di Kalangan Guru Ictl. *Pembinaan Media Pengajaran Berasaskan Multimedia Di Kalangan Guru ICTL*, *Bil 1*(2), 20–31.

[20] Mohammad Noor, N. A., Mahamod, Z., Hamat, A., & Embi, M. A. (2012). Persepsi pelajar terhadap aplikasi perisian multimedia dalam pembelajaran komsas bahasa melayu tingkatan 1. *Jurnal Pendidikan Bahasa Melayu*, *2*(1), 1–16.

[21] Embong, H. (2005). Pembinaan dan keberkesanan penggunaan perisian multimedia elektrokimia dalam pengajaran dan pembelajaran kimia tingkatan 4, 1–162.

[22] Mohd Yusoff, N., Osman, R., Shaari, A. S., & Ghazali, M. I. (2012). Kemahiran Belajar dalam Kalangan Pelajar Bidang Pendidikan. *Seminar Kebangsaan Majlis Dekan Pendidikan IPTA 2012*, 1–12.

[23] Richey, R. C., & Klein, J. D. (2014). Design and development research: Methods, strategies, and issues. Routledge.

[24] Hat, N. C., Sha'ari, S. H., & Abdul Hamid, M. F. (2013). Persepsi pelajar terhadap penggunaan animasi dalam pembelajaran bahasa arab. *Jurnal Teknologi (Sciences and Engineering)*, *63*(1), 25–29.

[25] Sardone, N. B., & Devlin-Scherer, R. (2010). Teacher candidate responses to digital games: 21st-century skills development. *Journal of Research on Technology in Education*, *42*(4), 409-425.

[26] Agustini, P. P., & Kristiantari, M. G. R. (2016). Penerapan Model Pembelajaran Berbasis Masalah Berbantuan Media Audio Visual Untuk Meningkatkan Hasil Belajarketerampilan Menyimak Tema Sejarah Peradaban Indonesia Pada Siswa Kelas V Sdn 8 Sumerta.

[27] Zainal, Z. (2004). Pembangunan Perisisan Multimedia Berasaskan Kecerdasan Ruang Visual Dan Interpersonal Bagi Tajuk Koordinat Tingkatan 2.

[28] Teoh, B. S. P., & Tse-Kian, N. E. O. (2007). Interactive multimedia learning: Students' attitudes

and learning impact in an animation course. *TOJET: The Turkish Online Journal of Educational Technology*, *6*(4).

[29] Rashid, Z. A., Kadiman, S., Zulkifli, Z., Selamat, J., & Hashim, M. H. M. Review of Web-Based Learning in TVET: History, Advantages and Disadvantages.

IOP Conf. Series: Materials Science and Engineering **203** (2017) 012024 doi:10.1088/1757-899X/203/1/012024

Engineering Students: Enhancing Employability Skills through PBL

Othman, H.[1,*], Mat Daud, K. A.[2], Ewon, U.[3], Mohd Salleh, B.[4], Omar, N. H. [1], Abd Baser, J.[1], Ismail, M. E.[1] & Sulaiman, A.[6]

[1]Faculty of Technical Education, Universiti Tun Hussein Onn Malaysia, Johor, Malaysia
[2]Faculty of Creative Technology and Heritage, Universiti Malaysia Kelantan, Malaysia
[3]Unit Sports, Co-Curricular & Culture, Politeknik Kota Bharu, Kelantan, Malaysia
[4]Center for Language Studies, Universiti Tun Hussein Onn Malaysia, Johor, Malaysia
[6]Center for General Studies and Co-Curiculum, Universiti Tun Hussein Onn Malaysia, Johor, Malaysia

*hasyam@uthm.edu.my

Abstract. As a developing country, Malaysia faces challenging tasks to develop her economy just like many other countries. Nowadays, change involves many aspects like the economy from agriculture to manufacturing, technology from modern to more advanced ones; mindset from traditional to advanced and so on. Previous studies show that one of the major issues facing local graduates is the lack of employability skills. This problem concerns not only the government but undergraduates and institutions alike. From the pedagogical aspect, one of the more effective ways to improve this is through instructional delivery and in this case the use of Problem-based Learning (PBL). The need to adopt PBL should involved applied subjects undertaken by engineering students. Studies have shown that the use of PBL has been proven to make learning more attractive and effective. In this research, we studied the effectiveness of PBL towards enhancing employability skills among engineering undergraduates. This study adopted a combination of qualitative and quantitative approaches. Data was collected using documents analysis. Student samples comprised manufacturing engineering undergraduates from public institutions of higher learning in Malaysia. The results show that student's employability skills can be enhanced using PBL. In addition, students become more competitive towards making them more relevance with the needs of the industry with regard to employability skills. In conclusion, PBL is a very effective catalyst towards raising the employability skills among engineering undergraduates and should be adopted in all engineering education.

Keywords: PBL; Employability Skills; Learning model; Engineering; Engineering Education; Engineering Education Students

1. Introduction

Like any other developing nations, Malaysia also has her share of education challenges in engineering education. As a relatively young country, we try to bring change from one of agricultural to that of industrialized and developed. There are many negative comments about locally trained graduates who lack the crucial employability skills required by the industry. Education is undeniably one of the most important factors that contributes to the development of any country. And in order to be competitive in the modern economy Malaysia should reform her education towards a more competitive one. In Malaysia, every effort is being taken by the Ministry of Education (MOE) and the Ministry of Higher Education (MOHE) to improve and raise the education system towards producing competent manpower for the country [1] [11]. In many conventional education system especially at school level, teachers give as much guidance as possibly could while the learners merely play the role of passive recipients. With limited face-to-face classroom hours and what more the heavily exam-oriented education system not much time is allocated for active learner interaction among themselves and the teacher. Even though the conventional approach of merely listening to lectures can deliver but this approach has been criticised for producing students who are passive and do not take into account the different needs and abilities of the students [2] [9] [17]. In general, conventional learning merely requires students to go through rote learning where interaction, a most crucial element for effective learning, is absent. According to [11], there is clear evidence showing "mismatch" of skills required by the employer and the ability of the graduates. According to [4] [8], "The industry told us that we are teaching all the wrong things. They explained that the quality of our [locally trained] graduates is so bad that 100% of the incoming goods [they prodice] require rework – that is, more training". At university, it is important that students interact more often among themselves, with the lecturers and learning materials. [27] stressed the importance of students interacting frequently because study life at university or college is very demanding and requires high level interaction among the campus citizens and learning materials. Therefore, it is only approapriate that a better learning approach that provides more interaction and learning opportunity be adopted towards producing graduates with higher employability skills needed by the industry.

1.1. Research Problem

The economic scenario of Malaysia which was previously conventional has now changed to K-economy. Malaysia is now known as a New Industrialized Country (NIC) meaning we need fully and semi-skilled workforce who possess the necessary features of a k-worker as required by the industry. This is in tandem with the views of [12], President of the

Malaysian Workers Association, who said that today's graduates must not treat their certificates and diplomas as the sole collateral for securing suitable jobs. Instead, our graduates must equip themselves with generic skills like interpersonal, information and communication technology, entreprenuership, leadership and English language. In relation to that, [30] said that collaborative learning (CL) combined with Problem-based Learning (PBL) should be apllied in the classroom as it helps to encourage active learning among the students. Both PBL and CL are student-oriented learning approaches and have been proven successful in engaging students to focus and learn more deeply on related subject matters. According to [7] [18], learning should be meaningful and purposeful because both are important elements for students to be fully engaged in their learning process. [29] was of the view that to force oneself to work harder in a group is crucial towards effective management of the working environment which is getting more sophisticated and challenging and PBL can deliver this. University curriculum should not only emphasize on technical knowledge but communication and collaborative skills should be infused among the undergraduates. Students who work well in groups can handle, solve problems and apply better communication strategy related to their study or work better. They also need to be given ample opportunity for sharing of ideas towards forming alternative solutions in problem solving. According to [10] [14], the sharing of knowledge helps to form better learning environment that would be otherwise difficult. PBL is thus a learning approach that uses current issues. Other than being able to increase student's learning skills this approach also helps to expose students to real world problems that they would encounter at the workplace upon graduation later. In general, scholars are of the view that PBL is a group learning strategy whereby the students interact among gthemselves and help each other in the learning process. Using PBL, the different levels of learning abilities among the students (higher and low) can actually benefit both parties. The weaker students can get help from the better ones and at the same time the better ones can better consolidate their mastery of knowledge and skills. No matter what, both parties can also benefit in terms of development of generic skills like teamwork, leadership, communication skills, respect and deep understanding.

1.2. Statement

The [28] said that PBL has been proven to have the ability to deliver learning content effectively. This view is supported by [24], who felt that that PBL is an effective skills training method in problem soling at the work place. According to [23], PBL can help build the needed important realationship between theory, application, education as well as professional practice expected of completing graduates.

1.3. Objectives

This research embarks on the following objectives: To identify the effectiveness of using PBL in teaching and learning; To identify the necessary employability skills required of graduates for employment in the industry; To identify the relationship betwen PBL and employability skills; To develop a model for the infusion of employability skills based on PBL for use at public institutions of higher learning in Malaysia; To evaluate the PBL model of development.

1.4. Importance

PBL can be seen as an effective alternative to the current teaching and learning strategies used in the learning classrooms. The outcome of this study should be beneficial to all especially the students, lecturers, university and stakeholders.

2. Methodology

2.1. Theoretical Framework

The framework of this study comprise elements of Manufacturing Industry, Employability Skills, PBL, Manufacturing Students, Manual and Facilitators of Public Universities. The Manufacturing Students are those undertaking a bachelor degree programme at public universities in Malaysia as primer data that using Multiple Regression. Employability Skills in divide by three sub area, it is Fudamental Skills, Team Work Skills, and Self-Management Skills acording to [25] & [26]. The PBL element actually describes the learning process involved beginning with identifying problems, generation of facts and ideas, learning issues, independent learning, synthesis and application, reflection and feedback and the final solution. [16] [20] told that potential of PBL to support enhancement of employability skills among students. This research also study connection and function of PBL in student's employability skills. This is very crucial when they try to adapt to real life, especially industrialisation world. If they had enough employability skills and be competence to industry needs that will help them better and more competitive.

2.2. Method

This study adopted the use of qualitative and quantitative approaches involving document analysis such as document by [21], [22], [25], [26], [3] and [13]. The combination of views from the industry and public sectors should contribute towards the development of a more accurate and comprehensive PBL manual for enhancing employabililty skills. It also helped the researcher in identifying a suitable design and delivery methid for the maual to be developed. The research involved perusing the notes, syllabus and related learning materials for the purpose. As for sampling of quantitative approach, the samples of this study were collected from manufacturing undergraduates from Universiti Teknikal Malaysia Melaka (UTeM). This is using quasi-experimental approach.

IOP Conf. Series: Materials Science and Engineering **203** (2017) 012024 doi:10.1088/1757-899X/203/1/012024

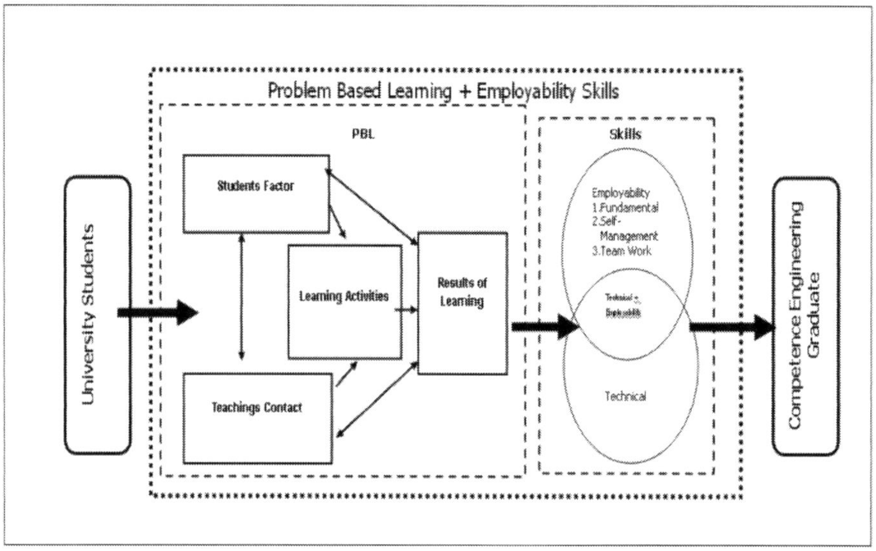

Figure 1: Adaptation Embeded Concept PBL 3P from [5], [16], [15], [21], [22], [25], [26], [3], and [13].

Table 1: The chronological of Research Methodology

No	Research Statements	Research Methodologies	Result of Research
1	What are the most important employability skills among the undergraduates?	Literature and Document Analysis	the most important employability skills among the undergraduates student in Malaysia
2	To what extent are elements of employability skills infused in the teaching and learning process of the undergraduates?		elements of employability skills infused in the teaching and learning process of the undergraduates student in Malaysia
3	Can PBL raise employability skills among the undergraduates?	Multiple Regression	PBL posilble or not to raised employability skills among the undergraduates.
4	Is PBL useful for enhancing employability skills?		PBL useful or not for enhance employability skills
5	What is the most suitable PBL model for enhancing employability skills?		suitable Model of PBL to enhance employability skills through undergraduates student in Malaysia

Table 2: Data Gathering Activities

Weeks	Activities	
	Treatment Group (PBL) 50 students	**Control Group (Conventional) 50 students**
1	Test 1 : Employability Skills 40 minutes	Test 1 : Employability Skills 40 minutes
1-10	Treatment Level 1 (PBL)	No Treatment Level 1 (Conventional)
11	Test 2 : Employability Skills 40 minutes	Test 2 : Employability Skills 40 minutes
1-16	Treatment Level 2 (PBL)	No Treatment Level 2 (Conventional)
20	Test 3 : Employability Skills 40 minutes	Test 3 : Employability Skills 40 minutes
40	Summative data as support from Industrial practical supervisors	

3. Results and Discussion

Results from study show the potential of PBL to be the alternative method for enhancing employability skills among Engineering Students. From document analysis researchers have proven what [10], [19], [25] & [26] have discovered and claimed, categories of important employability worth. This means in Malaysia employability skills can be divided into three sections; Fundamental Skills, Self-Management Skills, and Team Work Skills. Findings from Document analysis also found that important employability skills required by employers in Malaysia as in Table 3 and [5].

Table 3: Skills Needed among Graduates by the Employer for the Post of an Engineer, [5].

Skills	F	%
Effective communication skills	18	39.13
Interpersonal, personality skills	17	36.96
Able to work independently	13	28.26
Able to plan, manage, organize a group (teamwork)	12	26.09
Computer Literacy	12	26.09
Logical and strong analytical Skills	12	26.09
Resourceful and knowledgeable	10	21.74
Strong leadership qualities	7	15.22
Problem-solving skills	7	15.22
Self motivated	7	15.22
Dynamic, enthusiasm, aggressive, and energetic	6	13.04
Proactive, initiative, and creative	5	10.87
Honesty, integrity and commitment	5	10.87
Good presentation (report writing) skills	4	8.70
Able to work under pressure (tight schedule) with minimum supervision	4	8.70
Fast learner, quick learner, adaptability	4	8.70
Output or result oriented	4	8.70
Professionalism	4	8.70
Able to supervise a group	2	4.35
Responsibility	1	2.17
Discipline	1	2.17
Strategic thinking abilities	1	2.17

Table 3 also shows other results obtained from document analysis concerning the important employability skills infused in PBL process. From the quasi-experimental study involving manufacturing engineering students at UTeM, researchers found that student employability skills improved better compared to the conventional or control group. This shows the possibility and effectiveness of PBL on engineering subject that can enhance employability skills. This is shown in Figure 2. Meanwhile, Figure 3 shows that increase of employability skills in Team Work skills was up to 56%, followed by Fundamental Skills up to 30% and up to 14% for Self-Management Skills.

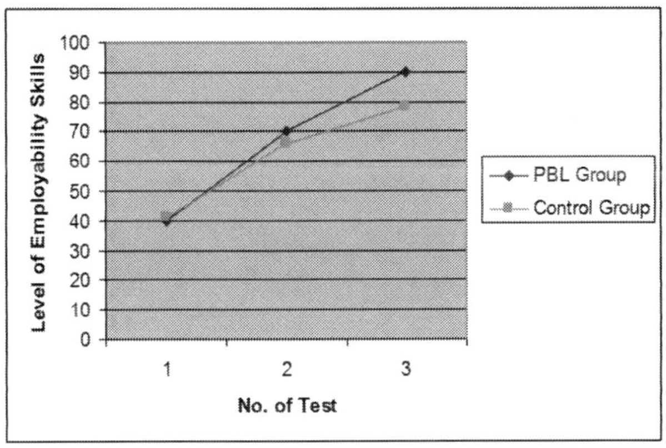

Figure 2: Employability Skills Level versus Employability Skills Test

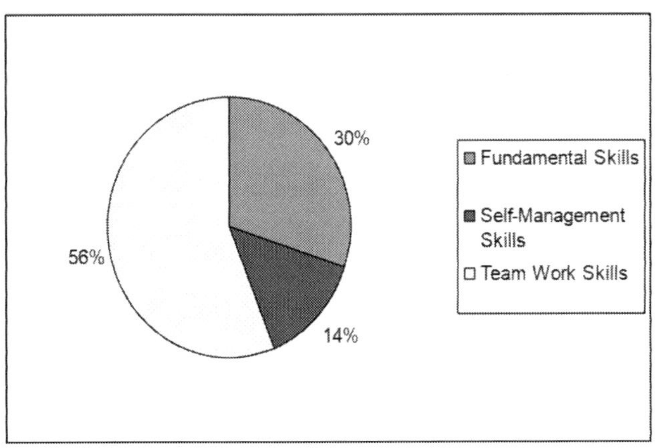

Figure 3: Employability Skills Categories

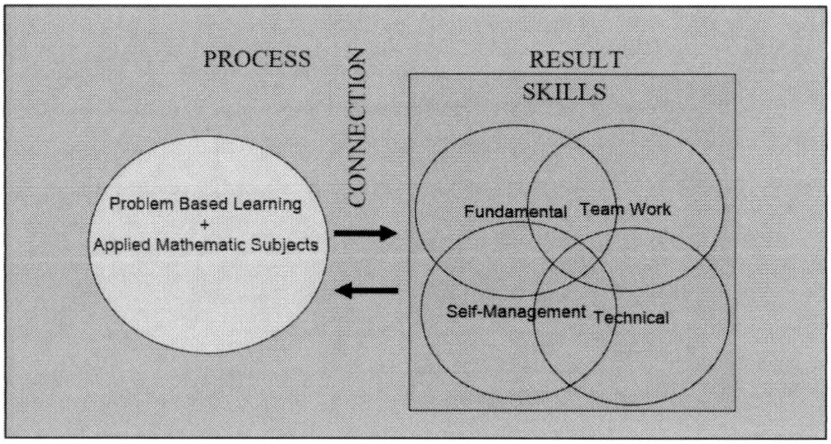

Figure 4: Model Connection of Employability Skills with Engineering Subject using PBL Methodologies

4. Conclusion

It could be summarized that unlike the conventional learning approaches of merely lectures that has many limitations and restrictions for effective learning using PBL has helped the students to become more independent and active in their learning process. This is prove that PBL to Engineering Students can make better student employability skills. This is what industries need and student are develop a lot of competence to survive in real life. They also had that much opportunity to develop their employability skills like communication, leadership and problem solving. In this study, with one subject using PBL students performed better than the conventional group. Now we can ask ourselves wouldn't it be better if we could allow all our students to learn using PBL. This study also cannot deny the possibility that PBL is the better alternative learning process. Conclusion can be made that this PBL model for Engineering Subjects have better connection, infusion with employability skills as depicted in Figure 4. For future study, the researchers would like to suggest 1) a qualitative study using PBL to gain in-depth views in this area; 2) a study to identify the best PBL model among the various models for Engineering and 3) a study to identify any other potentials of PBL in the learning processes. Finally, PBL has not only helped the students but helped the lectures to provide a more systematic student-centered learning environment towards producing better employable graduates. In conclusion, learning Engineering using PBL can better enhance Employability Skills.

Acknowledgement

The authors would like to acknowledge all the PBL Task Force members of UTHM for their continuous support and commitment in ensuring the success of our PBL project. Our thanks are also dedicated to UTHM, UTM, Fakulti Pendidikan Teknikal dan Vokasional, UTHM, and FKP UTeM for making this PBL research reality.

References

[1] Ahmad, I., Aris, B. dan Harun, J. (2005). Pembelajaran Mata Pelajaran ICT menggunakan ICT menerusi pendekatan PBM. Prosiding Seminar Pendidikan JPPG 2005, 1 (1), 271–276.

[2] Almajed, A. , Skinner, V. , Peterson, R. , & Winning, T. (2016). Collaborative Learning: Students' Perspectives on How Learning Happens. Interdisciplinary Journal of Problem-Based Learning, 10*(2)*.

[3] Australian Chamber of Commerce and Industry and Business Council of Australia (2002). Employability skill for the future. Technical Report, Department of education, Science and Training, Canberra.

[4] Badaway, M. K. (1995). Developing Managerial Skills in Engineers and Scientists: Succeeding As a Technical Manager. Van Nostrand Reinhold.

[5] Bakar, A.R., Mohamed, S., and Hanafi, I. (2007). Employability Skills: Malaysian Employers Perspectives. The International Journal of Interdisciplinary Social Sciences, 2 (1), 263–274.

[6] Beaumont C., Sackville A. and Cheng, C. (2003). Identifying Good Practice in the Use of PBL to Teach Computing. Polytechnic, Singapore.

[7] Belland, B. R., Burdo, R., & Gu, J. (2015). A blended professional development program to help a teacher learn to provide one-to-one scaffolding. Journal of Science Teacher Edu. www.ijpbl.org (ISSN 1541-5015) September 2016 | Volume 10 | Issue 2 cation, 26(3), 263–289. http://dx.doi.org/10.1007/s10972 -015-9419-2

[8] Bridges, S. M., Green, J. Botelho, M. G., & Tsang, P. C. S. (2015). Blended learning and PBL: An interactional ethnographic approach to understanding knowledge construction in-situ. In A. Walker, H. Leary, C. E. Hmelo-Silver, & P. A. Ertmer (Eds.), Essential Readings in Problem-Based Learning: Exploring and Extending the Legacy of Howard S. Barrows (pp. 107–130). Illinois: Purdue University Press.

[9] Chan, L. K., Bridges, S. M., Doherty, I., Ng, M. L., Sharma, N., Chan, N. K., & Lai, H. Y. Y. (2015). A qualitative study on how health professional students and their PBL facilitators perceive the use of mobile devices during PBL. Interdisciplinary Journal of Problem-Based Learning, 9(1): 83–95.

[10] deChambeau, A. L. , & Ramlo, S. E. (2017). STEM High School Teachers' Views of Implementing PBL: An Investigation Using Anecdote Circles. Interdisciplinary Journal of Problem-Based Learning, 11*(1)*.

[11] Hesketh, A.C. (1999). Towards a new economic sociology of the student financial experience of higher education. Journal of Education Policy, 14 (4), 385–410.

[12] Idris, I. (2005). Negara perlukan lebih 40,000 tenaga professional. Retrieved from http://www.Laman Web Adun Negeri Johor.htm. on 05 March 2007.

[13] Kementerian Pengajian Tinggi (2006). Retrieved from on from http://www.mohe.gov.my 16 September 2007.

[14] Lippert, K.S. and Granger, J.M. (1997). Peer learning in an introductory programming course. In Proceeding of the 12th Annual Conference on International Academy for Information management, 123–130.

[15] Lynda, W.K.N. (2004). Jump Start Authentic PBL. Singapore: Prentice Hall, 22–131.
[16] Moesby, E., Jahannsen, H.H.W., and Kornov, L. (2006). Individual Activities As An Integrated Part of Project Work: An Innovative Approach To Project Oriented and Problem-Based Learning (POPBL). World Transactions on Engineering and Technology Education, 5, 11.
[17] Mok, S.S. (1997). Pedadogi 2: Pelaksanaan Pengajaran. Kuala Lumpur: Kumpulan Budiman Sdn Bhd.
[18] Parnell, D. (1995). Why do I have to learn this? Waco: Center for Occupational Research and Development.
[19] Savin-Baden, M. (2016). The Impact of Transdisciplinary Threshold Concepts on Student Engagement in Problem-Based Learning: A Conceptual Synthesis. Interdisciplinary Journal of Problem-Based Learning, 10*(2)*.
[20] Savin-Baden, M., Poulton, T., Beaumont, C., & Conradi, E. (2016). What is real? Using problem-based learning in virtual worlds. In S. Bridges, L. K. Chan, & C. E. Hmelo-Silver (Eds.), Educational technologies in medical and health sciences. New York: Springer
[21] Secretary's Commission on Achieving Necessary Skills (SCANS). (1991). What Work Requires of Schools: A SCANS Report For America 2000. Washington DC: Department of Labor.
[22] Secretary's Commission on Achieving Necessary Skills (SCANS). (1992). Learning a Living: A blueprint for high performance: A SCANS report for America 2000. Washington, DC: U.S. Department of Labor.
[23] Steinemann, A. (2003). Implementing Sustainable Development Through Problem-Based Learning: Pedagogy and practice. Journal of professional issues in Engineering education and practices, 129 (4), 216–224.
[24] Stepien, W. J., Gallager, S. A., and Workman, D. (1993). Problem Based Learning for Traditional and Interdisciplinary Classrooms. Journal for the Education of the Gifted, 4, 338–345.
[25] The Conference Board of Canada. (1996a). Yukon work future: Skills for today's workplace. Ottawa, Ontario: The Corporate Council on Education, a program of the National Business and Education Center.
[26] The Conference board of Canada. (1996b). Employability Skills Profile: The Critical Skills Required of The Canadian Workforce. Retrieved. from http://www.conferenceboard.ca/libraries/educ_public/emskill.sflb
[27] Thomas, S.L. (2000). A social Network Approach to Understanding Students Integration Persistence. The Journal of Higher Education, 71 (5), 46–52.
[28] Ward, J.D and Lee, L.L. (2004). Teaching Strategies for FCS: Student Achievement in problem-Based Learning Versus Lecture-Based Instruction." Journal of Family and Consumer Sciences, 96 (1) 23–32.
[29] Yazici, J.H. (2005). A study of collaborative learning style and team learning performance. Journal of Education and Training, 47 (3), 216–229.
[30] Yerion, A.K. dan Rinehart A.J. (1995). Guidelines for collaborative learning in computer science. Bulletin of SIGCSE, 27 (4), 29–34.

IOP Conf. Series: Materials Science and Engineering **203** (2017) 012025 doi:10.1088/1757-899X/203/1/012025

Effective moisture diffusivity and activation energy of rambutan seed under different drying methods to promote storage stability

So'bah Ahmad[1a], Mohd Shamsul Anuar[1b], Farah Saleena Taip[1c], Rosnah Shamsudin[1d], Siti Roha A.M[2]

[1]Department of Process and Food Engineering, Faculty of Engineering, Universiti Putra Malaysia, 43400 UPM Serdang, Selangor
[2]Food Technology Programme, School of Industrial Technology, Faculty of Applied Sciences, UiTM 40450 Shah Alam, Selangor

Email: mshamsul@upm.edu.my

Abstract. The effects of two drying methods, oven and microwave drying on the effective moisture diffusivity and activation energy of rambutan seed were studied. Effective moisture diffusivity and activation energy are the main indicators used for moisture movement within the material. Hence, it is beneficial to determine an appropriate drying method to attain a final moisture content of rambutan seed that potentially could be used as secondary sources in the industry. An appropriate final moisture content will provide better storage stability that can extend the lifespan of the rambutan seed. The rambutan seeds were dried with two drying methods (oven and microwave) at two level of the process variables (oven temperature; 40°C and 60°C and microwave power; 250W and 1000W) at constant initial moisture contents. The result showed that a higher value of effective moisture diffusivity and less activation energy were observed in microwave drying compared to oven drying. This finding portrays microwave drying expedites the moisture removal to achieve the required final moisture content and the most appropriate drying method for longer storage stability for rambutan seed. With respect to the process variables; higher oven temperatures and lower microwave powers also exhibit similar trends. Hopefully, this study would provide a baseline data to determine an appropriate drying method for longer storage period for turning waste to by-products.

1. Introduction

An increasing number of world hunger is mainly due to 40-50% in food loss and food waste and in turn, has led to a dire requirement in drying process improvement. Drying process needs as a tool for prolonging the lifespan of food waste that is believed to have potential to be used as an alternative food resource (FAO, 2015). Food loss expresses as a food source degradation in term of quality and quantity due to the production process. While food waste is a non-food product that has the potential to be used in another industry. Though, sourced that could be defined as food loss and food waste must be from agriculture and fisheries bases. The importance of the food waste as the main subject to be studied, due to its ability to reduce the hunger percentage via providing a sustainable resource in the food system, remain in the eco-nature environment and foster in nutritional food as well as food security (FAO, 2015).

The simplest process in converting food waste to by-product that could be used to industry and further realization of all diversifying functions of food waste is drying. Drying is moisture removal process to avoid microorganism multiplication that could lead in shifting the chemical and physical structure of the material (Geankoplis, 2003). This process is seen vital for food waste that is seasonal in nature such as rambutan seed as the subject in this study.

Rambutan seed used as the subject in this study due to its diversity functions as an alternative to cocoa butter in chocolate manufacturing (Issara, Zzaman, & Yang, 2014), potential fat extracted as cosmetics (Lourith, Kanlayavattanakul, Mongkonpaibool, Butsaratrakool, & Chinmuang, 2016) as well as beneficial for medicinal purposes where rambutan seed extract has the potential to react as αglycosidase which helps to reduce the risk sugar forming in blood (Sylvia Soeng et al., 2015). In addition, the balanced content of carbohydrates, protein, fiber, vitamins, minerals and fatty acids that are needed in the body cause its potential as a functional food product as Rambuseed Florentine (Hussin, 2015). Furthermore, the rambutan seed has been used as a snack food among the people of Java, Indonesia since ancient time. Hence, food waste classification could probably consider depending on cultural and lifestyle of the origin country. However, the various cultural and lifestyle have an advantage in developing research mainly on rambutan seed. This is because based on the unique dieting of rambutan seed as it is reliably high in nutrients content and effectively reducing the increase of glucose in blood among the Indonesians has led to ongoing research to prove through scientific finding (Slyvia Soeng, Evacuasiany, Widowati, & Fauziah, 2015; Sylvia Soeng et al., 2015; Solís-Fuentes, Camey-Ortíz, Hernández-Medel, Pérez-Mendoza, & Durán-de-Bazúa, 2010). Nevertheless, due to its seasonal nature, drying process that could prolong the lifespan of rambutan seed without changing the quality, structure and chemical properties is critically needed.

The efficiency drying process is provided through an effective use in term of time, energy and cost. Time effectiveness in the drying process can also be seen through the speed of moisture removal. The moisture removal is directly dependent on the drying method applied. Drying method will affect the technique of moisture movement toward evaporation for the drying process. Moisture movement also depends on the effective moisture diffusivity and activation energy.

Effective moisture diffusivity is defined as moisture movement and closely related to drying rate. These two are the difference as the effective moisture diffusivity is relatively related to moisture velocity within the material, while the drying rate is the moisture vaporizing rate to air or simply a conversion rate of moisture to vapor by evaporation and depend directly on the pressure gradient that exists between material and the air due to a temperature gradient. Meanwhile, effective moisture diffusivity depends directly on the moisture gradient within the material, material structure, and drying method. Thus, the effective moisture diffusivity is among the most important parameters to determine an appropriate drying method to a substance because it helps the moisture movement within the material to facilitate the drying process and thereby achieve equilibrium moisture content or final moisture content that has been set for a substance. Therefore, indirectly, effective moisture diffusivity is the parameter that influences the drying rate of a substance as well as the indicator to determine an appropriate drying method that could extend the product lifespan. Besides, effective moisture diffusivity is also closely related to the activation energy.

Activation energy is moisture particles bonding and needs to be severed for moisture movement. The activation energy is inversely proportional to the effective moisture diffusivity where low activation energy will refer to less energy needed to sever moisture particle bonding and accelerates moisture velocity and provide a higher value of effective moisture diffusivity. This will help to speed up moisture removal process to the environment and to shorten the drying time and promote an appropriate drying process in terms of time and energy consumption. Whereas, if the activation energy is high, the energy requirement to split moisture particle bonding is higher and lessen in moisture velocity and will provide a lower value of effective moisture diffusivity and require a longer drying time.

Evaluation of effective moisture diffusivity and activation energy is often associated with drying characteristics and the main focus previously was based on the effect of drying condition on effective moisture diffusivity and activation energy of product (Aghbashlo, kianmehr, & Samimi-Akhijahani, 2008; Azzouz, Guizani, Jomaa, & Belghith, 2002; Babalis & Belessiotis, 2004; Motevali, Abbaszadeh, Minaei, Khoshtaghaza, & Ghobadian, 2012).It is clearly illustrated via a diversity of foodstuffs that have been studied and have been formulated that 92% of food product fall in effective diffusivity range between 10^{-12} to 10^{-8} m^2s^{-1} (Zogzas, Maroulis, & Marinos-Kouris, 1996). Based on the combination of the author's knowledge, literature based on the interrelationships between the effect of drying method, process variable (Zogzas et al., 1996) and material properties (Xiong, Narsimhan, & Okos, 1991), the effective moisture diffusivity and activation energy have been studied. Conversely, to date, the relationship of effective moisture diffusivity and activation energy on the equilibrium moisture content of rambutan seed as a food waste at two different drying methods; microwave and oven and relation to prolong the lifespan of rambutan seed and promote storage stability is still vague. Hence, there is a need to study the direct relationship between the effective moisture diffusivity and activation energy in predicting an appropriate drying method to achieve an equilibrium moisture content and final moisture content of rambutan seed that led to prolonging lifespan and promote storage stability.

Extending the lifespan of food waste such as rambutan seed is needed as an alternative to save food. This is because, as the ability in the accessibility throughout the year, it will facilitate the conversion of rambutan seed to secondary sources to the industry, thus reducing waste and encourage recycling and reuse of waste to help preserve the environment as well as promoting a sustainable food chain and zero food waste.

The objective of this study is to evaluate the effective moisture diffusivity and activation energy at two drying methods; microwave and oven to determine an appropriate drying method of rambutan seed to achieve an equilibrium moisture content and final moisture content as a tool to extend shelf life and thus provide storage stability. Effect of process variable on effective moisture diffusivity was also reviewed to strengthen the justification for identifying an appropriate drying method to dry rambutan seed. Thus, by knowing the appropriate drying method to prolong the lifespan of rambutan seed, the use of food waste in providing alternative food sources and secondary sources in the industry thus, maximizing the use of natural resources could be realized. Indirectly, it is expected to help address the problems of poverty and hunger in the world, climate changes as well as increasing awareness on how to react to global food loss and waste and ultimately save food.

2. Material and methods

2.1. Material

Rambutan fruits were obtained from University Agricultural Park, Universiti Putra Malaysia (UPM), Selangor. Harvested fruit was stored in zip-lock bag polyethylene at 8.5°C in cool room Faculty of Engineering, Universiti Putra Malaysia prior deseeding. Rambutan seed was manually deseeded, washed, and left at room temperature up to 4 hours to dry up surplus water at seed surface. As surplus water dried, the initial moisture content will be measured using a digital moisture analyzer (OHAUS, MB45, UK) with triplicates. The seed was weighted to (5.0 ± 0.01 g) and heated at 103 ± 0.1°C until a constant weight was achieved.

2.2. Drying

2.2.1. Oven drying

5 g of rambutan seed was dried at two different oven temperatures 40°C and 60°C using an automatic electric oven model OF-22GW (Jelotech, Korea) with duplication. Weight loss was continuously measured every 20 minutes interval using an analytical balance model AY220 (Shimadzu, Japan) until weight between two repeated weighing was less than 0.005 g.

2.2.2. Microwave drying

5 g of rambutan seed was dried at two different level of microwave power 250W and 1000W using commercial microwave oven model NN-C2003S (Panasonic, Malaysia). Weight loss was recorded every 5 minutes interval using an analytical balance model AY220 (Shimadzu, Japan) with precision 0.0001 g until mass do not change between the two weighing interval. Each run was dried in duplicates.

2.3. Effective moisture diffusivity and activation energy

Effective moisture diffusivity value for oven and microwave drying was determined due to the slope of Handerson and Pabis drying model as in Eq. (1) and (2);

$$MR = a\ exp(-kt) \tag{1}$$

$$k = \frac{\pi^2 D_{eff}}{4L^2} \tag{2}$$

where k is a slope from Handerson and Pabis equation, D_{eff} is effective moisture diffusivity (m^2s^{-1}), L is half the thickness of slab (m).

Meanwhile, the activation energy value for the oven and microwave drying was measured based on Eq. (3) and (4) (Dadalı, Apar, & Özbek, 2007; Zogzas & Maroulis, 1996);

$$D_{eff} = D_o exp\left(\frac{-E_a}{RT}\right) \tag{3}$$

where D_o is the pre-exponential factor of the Arrhenius equation (m^2s^{-1}), E_a is the activation energy (kJ mol^{-1}), R is the ideal gas constant (8.3143 kJ $mol^{-1}K^{-1}$) and T is the absolute temperature (K).

$$D_{eff} = D_0 exp\frac{-E_a m}{P_m} \tag{4}$$

where D_o is the pre-exponential factor of the Arrhenius equation (m^2s^{-1}), E_a is the activation energy (Wg^{-1}), m is the weight of the raw material (g) and P is the microwave power (W).

3. Results and discussions

3.1. Effect of drying methods on effective moisture diffusivity

Figure 1 shows the effect of different drying methods on effective moisture diffusivity at two different intensity that represented by number 1 and 2; low and high intensity respectively. Effect of each method was examined at two level of intensity where low levels represent 40°C and 250W and higher levels represented by 60°C and 1000W.

Figure 1 shows that the microwave drying had a higher effective moisture diffusivity value compared to oven drying at low-level microwave power (250W) and oven temperature (40°C), portrayed by the two left vertical bars in figure 1 (labeled as 1). A higher value of effective moisture diffusivity shows faster moisture velocity in the material that will shorten the drying time to achieve the equilibrium moisture content or the targeted final moisture content. Reduction in the drying time indirectly, will reduce energy consumption and in turn, less processing cost. Therefore, microwave drying provides an appropriate drying technique in attaining the equilibrium moisture content (or the targeted final moisture content) and promote storage stability at the low level of oven temperature and microwave power. Less drying time in microwave drying may be due to the drying mechanism involved. A collision between the particles in the seed that leads to the increase in the kinetic energy and in turn generating heat internally accelerating the moisture velocity in the seed. The inherent structure of rambutan seed where the moisture is located in the deepest layers and having low moisture in general, hence, the microwave drying mechanism that heats internally is the most appropriate drying method to accelerate the moisture velocity in the seed towards the seed surface for the drying process to proceed via evaporation. Moisture velocity based on the moisture gradient also fasten moisture velocity and thus shorten the drying time to achieve an equilibrium moisture content or the targeted final moisture content.

However, the dissimilar trend can be observed at higher levels, where oven drying had a higher value of effective moisture diffusivity compared to microwave drying represented by the two right vertical bars in figure 1 (labeled as 2). This may probably due to drying at 1000W happened quickly and 135 minutes of total drying exposure is too long. A higher microwave power leads to huge moisture transfer at the beginning and as the drying proceeded moisture velocity becomes slower due to decrease level in moisture content and total moisture within the seed. Slower moisture velocity would prolong the drying time, lower the drying rate and in turn lessen the value of effective moisture diffusivity. A similar finding has also been observed for the drying of okra, whereas drying proceeded, lower moisture content removed due to a decrease in microwave absorption within the sample. Lower drying rate leading to lower effective moisture diffusivity (Dadalı, Kılıç Apar, & Özbek, 2007). It is suggested if higher microwave powers are required, shorter drying times are recommended.

But, if compared to these two drying methods, microwave drying at lower microwave power, 250 W would be an appropriate drying condition for the rambutan seed to prolong lifespan and longer storage stability. It gives a higher effective diffusivity of moisture, 3.13 x 10^{-7} m^2s^{-1}, shorter drying time thus saving energy and cost.

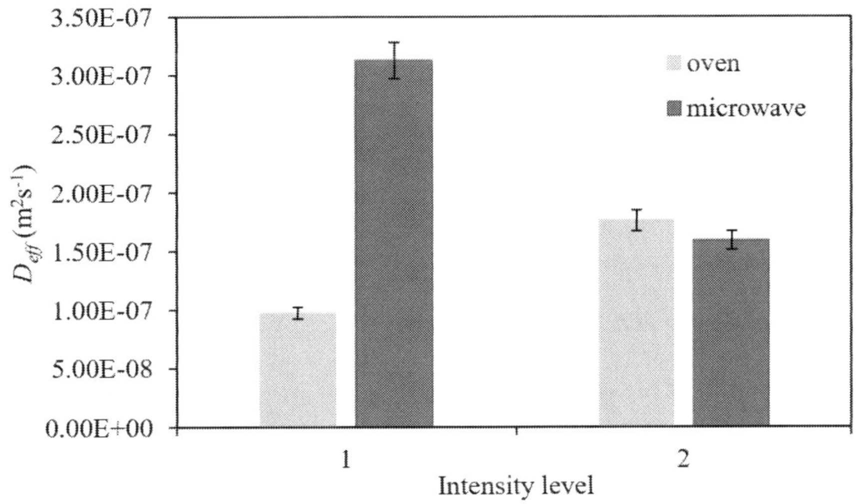

Figure 1: Effect of drying methods on effective moisture diffusivity at two level of intensity 1 and 2; low and high intensity, respectively.

3.2. Estimation of activation energy for each drying method

A higher activation energy represents a higher energy required to break moisture bonding during the drying process. Therefore, a higher activation energy will retard the moisture velocity and may probably give the lower amount of effective moisture diffusivity and thus prolongs the drying process. The value of D_o and E_a in oven drying were respectively estimated as 6.80 x 10^{-4} m^2s^{-1} and 22966.4 kJ mol^{-1} as showed in table 1.

Whereas, the value D_o and E_a were respectively estimated as 1.26 x 10^{-7} m^2s^{-1} and 45.05 Wg^{-1} for the microwave drying as presented in table 1. As activation energy in oven drying is dependent upon the oven temperature and microwave drying is based on the ratio of mass to microwave power applied, then, the activation energy is an unsuitable parameter to be used in comparing these two drying methods in determining an appropriate drying method of rambutan seed. Moreover, as a dissimilar mechanism was applied to split the moisture bonding during microwave and oven drying, thus, activation energy had turn to inappropriate parameter to be used for evaluating drying efficiency within these two drying methods. However, this parameter could be effectively used in comparing within similar drying methods at varied level of process variable for evaluating the effect of process variables on responses to attain an effective drying condition.

Table 1: Estimation of the activation energy for each drying method

	OVEN		MICROWAVE	
D_o	6.80 x 10^{-4}	m^2s^{-1}	1.26 x 10^{-7}	m^2s^{-1}
E_a	22966.4	$kJmol^{-1}$	45.05	Wg^{-1}

3.3. Effect of operational process variable on effective moisture diffusivity

3.3.1. Oven temperature

Figure 2 (a) shows that a higher oven temperature has higher effective moisture diffusivity compared to low oven temperature. A higher effective moisture diffusivity shows rapid moisture velocity and would shorten the drying time to achieve an equilibrium moisture content or the targeted final moisture content. A high temperature leading to a larger value of effective moisture diffusivity most probably due to the higher temperature gradient between seed surface and the air. Moisture removal in the oven drying is based upon the temperature gradient that acting as a driving force for moisture evaporation in the drying process. Therefore, high temperature would provide a greater driving force due to the larger temperature gradient that will expedite the moisture velocity from inner seed to the seed surface for evaporation to occur. Faster moisture velocity will decrease the drying time. As the drying time decreases, energy consumption will be reduced and in turn, would provide lower drying cost and promote an effective drying method for storage stability. Thus, 60°C in oven drying would be an appropriate drying condition to dry rambutan seed for storage stability compared the lower drying temperature of 40°C. The value for the effective moisture diffusivity for 40°C and 60°C are 9.74 x 10^{-8} m^2 s^{-1} and 1.76 x 10^{-7} m^2s^{-1} respectively.

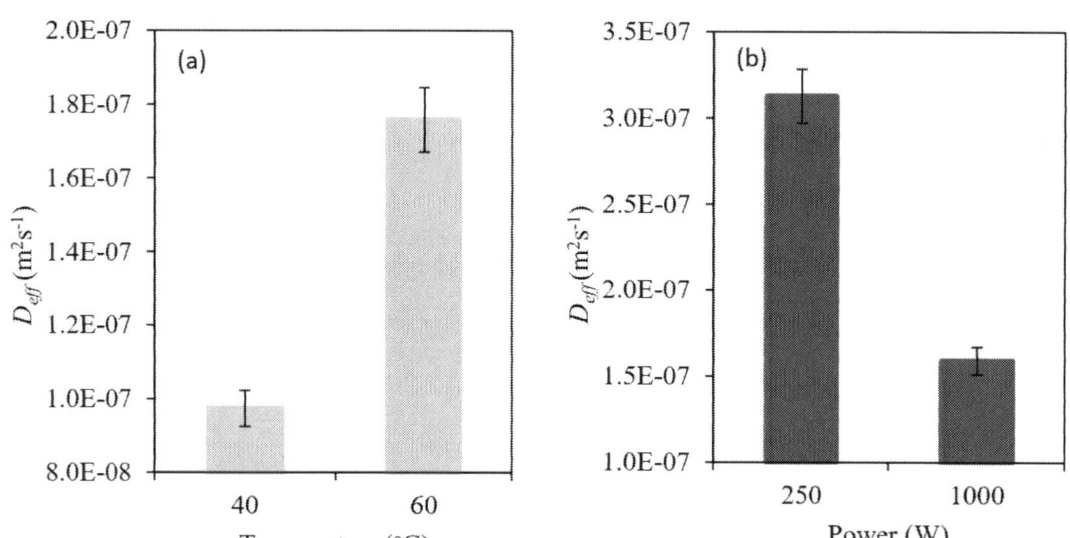

Figure 2: Effect of operational process variable on effective moisture diffusivity (a) Effect of oven temperature; (b) Effect of microwave power

3.3.2. Microwave power

Figure 2 (b) shows a low microwave power exhibit a higher effective moisture diffusivity for the rambutan seed compared to a high microwave power. This may probably due to drying mechanism in the microwave is based on the moisture gradient and the heat generated internally to accelerate the drying process. Therefore, a lower microwave power is required for moderate drying exposure or a short drying exposure if dried at higher microwave power, 1000W as applied in this study. A longer drying exposure at a high microwave power will reduce the drying rate and slow the moisture velocity within the material providing a lower value of effective moisture diffusivity. This may probably occur due to drying using a high microwave power will accelerate the moisture velocity and shorten the drying time. However, when the drying time is extended to achieve an equilibrium moisture content, the moisture content will be too low and the process for moisture removal as well as moisture velocity are too slow due to at this stage, the moisture often locates in the innermost layer of the seed. As moisture velocity is slow, a longer drying time and more energy are required that would lead to increase the drying cost. Therefore, with respect to microwave power, a lower microwave power, 250W become an appropriate drying condition for rambutan seed to prolong the lifespan. The value of effective moisture diffusivity for both microwave power of 250 and 1000 W is 3.13 x 10^{-7} m^2s^{-1} and 1.59 x10^{-7} m^2s^{-1}, respectively.

4. Conclusions

Based on this study, a high amount of effective moisture diffusivity was obtained at a high oven temperature (60°C) and a low microwave power (250W); 1.76 x 10^{-7} m^2s^{-1} and 3.13 x 10^{-7} m^2s^{-1}, respectively. A higher value of effective moisture diffusivity would accelerate moisture velocity within rambutan seed to facilitate moisture removal to achieve equilibrium moisture content at specified relative humidity. Therefore, a higher value of effective moisture diffusivity would be an appropriate parameter to design an effective drying method in terms of time as well as savings in the energy consumption and cost to prolong the lifespan of rambutan seed and promoting its storage stability. With respect to both drying methods applied and the range studied, a low microwave power (250W) become an appropriate drying method for rambutan seed to prolong its lifespan and storage stability.

Acknowledgements

The authors gratefully acknowledge Universiti Putra Malaysia (UPM) for IPS research grant (vote no. 9464600) and Ministry of Higher Education as well as Applied Science Faculty, UiTM, Shah Alam for financial sponsored throughout this study.

References

Aghbashlo, M., kianmehr, M. H., & Samimi-Akhijahani, H. (2008). Influence of drying conditions on the effective moisture diffusivity, energy of activation and energy consumption during the thin-layer drying of berberis fruit (Berberidaceae). *Energy Conversion and Management*, *49*(10), 2865–2871. https://doi.org/10.1016/j.enconman.2008.03.009

Azzouz, S., Guizani, A., Jomaa, W., & Belghith, A. (2002). Moisture diffusivity and drying kinetic equation of convective drying of grapes. *Journal of Food Engineering*, *55*(4), 323–330. https://doi.org/10.1016/S0260-8774(02)00109-7

Babalis, S. J., & Belessiotis, V. G. (2004). Influence of the drying conditions on the drying constants and moisture diffusivity during the thin-layer drying of figs. *Journal of Food Engineering*, *65*(3), 449–458. https://doi.org/10.1016/j.jfoodeng.2004.02.005

Dadalı, G., Apar, D. K., & Özbek, B. (2007). Estimation of Effective Moisture Diffusivity of Okra for Microwave Drying. *Drying Technology*, *25*(9), 1445–1450. https://doi.org/10.1080/07373930701536767

Dadalı, G., Kılıç Apar, D., & Özbek, B. (2007). Microwave drying kinetics of okra. *Drying Technology*, *25*(5), 917–924. https://doi.org/10.1080/07373930701372296

FAO. (2015). Global initiative on food loss and food waste reduction. *United Nations*, 1–8. Retrieved from http://www.fao.org/save-food/en/

Geankoplis, J. C. (2003). Drying of Process Materials. In *Transport Processes and Unit Operations*

(Fourth, pp. 559–624). New Jersey: Prentice Hall.

Hussin, M. (2015, October). Snek biji rambutan. *Agro, Harian Metro*, p. 2. Malaysia.

Issara, U., Zzaman, W., & Yang, T. A. (2014). Rambutan seed fat as a potential source of cocoa butter substitute in the confectionary product. *International Food Research Journal, 21*(1), 25–31.

Lourith, N., Kanlayavattanakul, M., Mongkonpaibool, K., Butsaratrakool, T., & Chinmuang, T. (2016). Rambutan seed as a new promising unconventional source of specialty fat for cosmetics. *Industrial Crops and Products, 83*, 149–154. https://doi.org/10.1016/j.indcrop.2015.12.045

Motevali, A., Abbaszadeh, A., Minaei, S., Khoshtaghaza, M. H., & Ghobadian, B. (2012). Effective Moisture diffusivity, activation energy and energy consumption in the thin-layer drying of jujube (Zizyphus jujube Mill). *Journal of Agricultural Science and Technology, 14*(3), 523–532.

Soeng, S., Evacuasiany, E., Widowati, W., & Fauziah, N. (2015). Antioxidant and hypoglycemic activities of extract and fractions of Rambutan seeds (Nephelium lappaceum L.). *Biomedical Engineering, 1*(1), 13–18.

Soeng, S., Evacuasiany, E., Widowati, W., Fauziah, N., Manik, V., & Maesaroh, M. (2015). Inhibitory potential of rambutan seeds extracts and fractions on adipogenesis in 3T3-L1 cell line. *Journal of Experimental and Integrative Medicine, 5*(1), 55–60. https://doi.org/10.5455/jeim.200115.or.120

Solís-Fuentes, J. a., Camey-Ortíz, G., Hernández-Medel, M. D. R., Pérez-Mendoza, F., & Durán-de-Bazúa, C. (2010). Composition, phase behavior and thermal stability of natural edible fat from rambutan (Nephelium lappaceum L.) seed. *Bioresource Technology, 101*(2), 799–803. https://doi.org/10.1016/j.biortech.2009.08.031

Xiong, X., Narsimhan, G., & Okos, M. R. (1991). Effect of composition and pore structure on binding energy and effective diffusivity of moisture in porous foods. *Journal of Food Engineering, 15*, 187–208.

Zogzas, N. ., Maroulis, Z. B., & Marinos-Kouris, D. (1996). Moisture diffusivity data compilation in foodstuffs. *Drying Technology, 14*(10), 2225–2253. https://doi.org/10.1080/07373930701438592

Zogzas, N. P., & Maroulis, Z. B. (1996). Effective moisture diffusivity estimation from drying data. A comparison between various methods of analysis. *Drying Technology, 14*(7 & 8), 1543–1573. https://doi.org/10.1080/07373939608917163

A Study on Shelf Life Prolonging Process of Chili Soy Sauce in Malaysian SMEs' (Small Medium Enterprise)

Zainon Binti Mat Sharif[1], Norhasnina Binti Mohd Taib[2], Mohd Sallehuddin Bin Yusof[3], Mohammad Zulafif Bin Rahim[4], Abdul Latif Bin Mohd Tobi[5], Mohd Syafiq Bin Othman[6]

[1,2]Graduate Business School, College of Graduate Studies, Universiti Tenaga Nasional, Putrajaya Campus, Jalan IKRAM-UNITEN, Kajang, Selangor, Malaysia.

[3,4,5]Department of Manufacturing And Industrial Engineering, Universiti Tun Hussein Onn, Parit Raja, 86400 Batu Pahat, Johor, Malaysia

[6]Faculty Hotel and Tourism, Universiti Teknologi Mara, Bertam Campus, 13200 Pulau Pinang, Penang, Malaysia.

zainon@uniten.edu.my, norhasninamohdtaib@yahoo.com, mdsalleh@uthm.edu.my, zulafif@uthm.edu.my, abdlatif@uthm.edu.my, mohdsyafiqq@gmail.com

Abstract. This research paper presents the possible solutions to prolong the shelf life of spicy (chili) soy sauce. The current spicy soy sauce formulation is without adding preservative which result in shorter shelf life. It is suggested to add chemical preservative to this spicy soy sauce in order to prolong its shelf life without jeopardising its prevailing taste. The proposed preservative is sodium benzoate. It is hope that by adding sodium benzoate, it can prolong the shelf life of the products from one year to two years without jeopardising the taste and quality of the products. The problem to extend the shelf life of spicy (chilli) soy sauce was 100% solved. The product could be extended to 2 years without adding any preservative (sodium benzoate) as the main raw material (soy sauce) purchased from "Kicap Jalen" had been added sodium benzoate as their preservative to prolong the soy sauce shelf life. All the physicochemical and nutritional analysis shown good results. As for the microbiological analysis, all the 3 samples shown good results on the total plate count.

Keywords: spicy, soy sauce, shelf life, sodium benzoate

1. Introduction

Soybeans have been consumed by Asian people in different types of traditional soy food products since 1000 years ago. There are many types of soy foods available on the market in eastern and western countries. Generally, soy foods are presented into two forms which are no fermented and fermented. Traditional no fermented soy foods include fresh green soybeans, whole dry soybeans, soymilk, tofu and whole-fat soy flour. The traditional fermented soy foods includes tempeh, miso soy sauce and fermented tofu. In addition, soybeans have been produced using both modern processing technique such as adopted by large soybean-processing plants and traditional ways such as oriental processing techniques [3].

Spicy (chili) soy sauce is a condiment prepared with red chili and premium soybeans sauce as a primary ingredient. Spicy soy sauce is spicy chili hot and sweet. It has rich soy aroma and hot flavor will delicately add excitement to any dish. This spicy soy sauce is best for dipping any crunchy crackers. Moreover, spicy soy sauce can be used as cooking glaze or used to marinade. This sauce is widely used in classic Asian especially Javanese people. Current formulation is without added preservative in spicy soy sauce, except for readymade soy sauce that normally added with some preservative. This spicy soy sauce are widely consumed amongst Malaysian, Brunei and Singaporean, making it highly demanding on market.

The making of spicy soy sauce involved lots of process and techniques. In addition, correct techniques and formulations will prolong shelf life, improve quality and safety of the products. However, microbial deterioration of spicy soy sauce is of major concern. Growth of moulds in spicy soy sauce results in changes in the organoleptic and microbiological which lead to spoilage. Moreover, the shelf life of current spicy soy sauce is only few months to one year. Hence, manufacturers is focusing on the options to prevent spoilage and quality changes during storage period.'

Development and improvement on the formulation of spicy soy sauce is crucial step in order to produce clean and holistic products as well as longer storage life and safe for human consumption. This is a very crucial step in food production especially on usage of safe ingredients especially food preservative. In order to ensure an economical, safe and quality product that last for longer period, addition of chemical preservative such as sodium benzoate is most likely used method. According to [5] and [1], sodium benzoate is very suitable for food that exhibits acidic condition in order to obtain microbial stability and high quality of the products. In addition, [4] addressed that, sodium benzoate is extensively popular as preservative in acidic food such as condiments, paste and salad dressing.

This conceptual paper proposes a simple application procedures, incurs low cost and safe chemical preservative formulation for organoleptically acceptable and shelf stable of spicy soy sauce.

Figure 1: Maji Chili Soya Sauce

2. Solution

2.1 Prolong the Shelf Life of Spicy Soy Sauce

Sodium benzoate has the chemical formula NaC7H5O2; it is a widely used in food preservative, with an E number of E211. It is the sodium salt of benzoic acid and exists in this form when dissolved in water. It can be produced by reacting sodium hydroxide with benzoic acid. Benzoic acid occurs naturally at low levels in cranberries, prunes, greengage plums, cinnamon, ripe cloves, and apples.

Sodium benzoate is a preservative. As a food additive, sodium benzoate has the E number E211. It is bacteriostatic and fungistatic under acidic conditions. It is most widely used in acidic foods such as salad dressings (vinegar), carbonated drinks (carbonic acid), jams and fruit juices (citric acid), pickles (vinegar), and condiments. It is also used as a preservative in medicines and cosmetics. Concentration as a preservative is limited by the FDA in the U.S. to 0.1% by weight. Sodium benzoate is also allowed as an animal food additive at up to 0.1% concentration according to AFCO's official publication.

2.2 Lab Testing

The researchers conducted some critical analysis for the company, namely the physicochemical, microbiological, organoleptic test and nutritional analysis of the spicy soy sauce. The test is very important to ensure the products saved to consume and can be accepted by the consumers. This analysis is very essential in order to determine the chemical composition of the existing sodium benzoate in the spicy soy sauce products. It will also provide the information related to the level of existing preservative and necessity to add extra sodium benzoate according to requirement by FDA in prolong the shelf life of spicy soy sauce.

The analysis samples include fresh spicy soy sauce product and also spicy soy sauce product with storage period of 1 year, 2 years, and 3 years old. All samples had been collected and send for analysis.

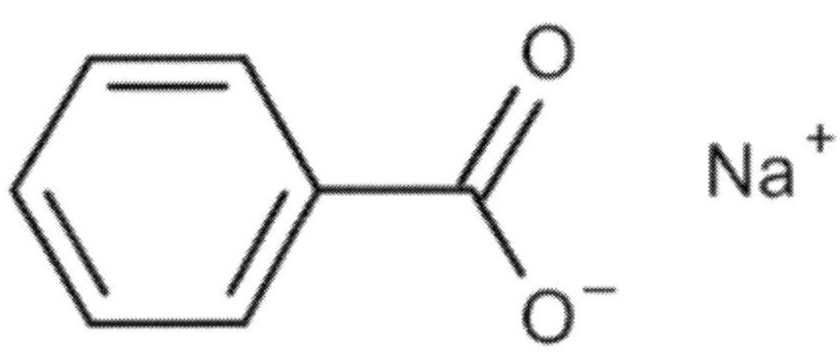

Figure 2: Chemical preservative (sodium benzoate)

3. Results

Prolong shelf life of spicy soy sauce

In this study, the spicy soy sauce will be added with sodium benzoate and analysed for microbiological, chemical composition and physical sensory. In this analysis, values of fatty acids, moisture content, soluble solid content, chemical composition and total phenolic content will be determined. Previous studies has shown that sodium benzoate can improve shelf life of food products and also provide stability in variety of food products with high organoleptic acceptability and stable nutritional quality. Once it is proven to be provide long shelf life, the final protocol will be recommended and to be followed by industry for shelf stable spicy soy sauce.

The researchers collected samples from Maji Enterprise to conduct laboratory critical analysis on the product. The analysis conducted to the products were physicochemical, nutritional analysis, and microbiological analysis. The samples collected included storage study from 2013, 2015, 2016 and 2017.

The problem to extend the shelf life of spicy (chilli) soy sauce was 100% solved. The product could be extended to 2 years without adding any preservative (sodium benzoate) as the main raw material (soy sauce) purchased from "Kicap Jalen" had been added sodium benzoate as their preservative to prolong the soy sauce shelf life. Results of the analysis was shown in Table 1.

Table 1, shown the full analysis that were done for spicy (chilli) soy sauce. The analysis shown that the shelf life for the product could be extended for 2 years without adding any preservative (sodium benzoate). This was because the main raw material (soy sauce) was supplied / purchased from "Kicap Jalen" and the existing raw material had been added sodium benzoate to extend shelf life of soy sauce.

All the physicochemical and nutritional analysis shown good results. As for the microbiological analysis, all the 3 samples stated expired 2017, 2016 and 2015 shown good results on the total plate count except sample stated expired 2013 was not in good condition as the results on total plate count was 1.39×10^3 cfu/g. Product quality and safety to consume by the consumer the total plate count should not be more or exceed than 1.00×10^4 cfu/g.

Shelf life is different from *expiration date*. Shelf life is linked to food quality, expiration date to food safety. A product that has passed its shelf life might still be safe, but quality is no longer guaranteed. In most food stores, shelf life is controlled by using stock rotation. This means moving products with the earliest sell by date to the front of the shelf, so that most shoppers will pick them up first and so getting them out of the store. This is important, as stores can be fined for selling out of date products. Most shops, if not all, will have to mark such products down as wasted, leading to a loss of profit.

Shelf life is can be changed by many things: exposure to light and heat, transmission of gases (including humidity), mechanical stresses, and contamination by things such as micro-organisms.

Mathematically, product quality is often modelled using only one parameter, for example the concentration of a chemical substance, a microbiological index, or a physical parameter. Sometimes, the parameter picked is irrelevant.

The shelf life is an important factor to health. Bacteria are everywhere, and foods left unused too long will often get large amounts of bacterial living in them. It may be dangerous to eat them and lead to food poisoning. The shelf life itself cannot always be trusted to tell how safe it is to eat a certain item of food. For example, pasteurized milk can remain fresh for five days after its sell-by date if it is refrigerated properly. In contrast, if milk already has harmful bacteria, the use-by dates do not matter.

IOP Conf. Series: Materials Science and Engineering **203** (2017) 012026 doi:10.1088/1757-899X/203/1/012026

Table 1: Results of Physico-chemical characteristics, Microbiological, Nutritional Aspects and Shelf life studies of Spicy (Chili) Soy Sauce

No	Parameters	Sample1 Expiry 7/17	Sample2 Expiry 2/16	Sample3 Expiry 10/15	Sample4 Expiry 9/13	Commercial label on the product	Range	Remarks
1	pH	4.32	4.16	3.99	3.85	NA	3.85 – 4.32	OK Acidic
2	Benzoic Acid - HPLC	254.73mg/kg	152.23mg/kg	315.31mg/kg	433.16mg/kg	NA	152 – 433mg/kg	OK Following FDA
3	Sorbic Acid	NA	NA	NA	NA	NA	NA	NA
4	Total Dietary Fibre	1.41g/100g	7.76g/100g	2.67g/100g	5.66g/100g	NA	1.41 – 7.76g/100g	OK FDA
5	Total Calories Energy Fat Protein Moisture Ash Carbohydrate	 251kcal/100g 1.17g/100g 2.76g/100g 32.04g/100g 7.40g/100g 56.64g/100g	 262kcal/100g 0.15g/100g 2.71g/100g 32.01g/100g 6.51g/100g 58.61g/100g	 245kcal/100g 0.14g/100g 1.69g/100g 33.44g/100g 6.84g/100g 57.90g/100g	 255kcal/100g 0.88g/100g 2.73g/100g 34.94g/100g 5.32g/100g 56.12g/100g	 236kcal/100g 0.2g/100g 1.6g/100g NA NA 56.9g/100g	 236 – 262kcal/100g 0.14 – 0.20g/100g 1.60 – 2.76g/100g 32.00 – 34.94g/100g 5.32 – 7.40g/100g 56.12 – 58.61g/100g	 OK OK OK OK OK OK
6	Total Plate Count	2.55 x 10^2 cfu/g	2.55 x 10^2 cfu/g	2.55 x 10^2 cfu/g	1.39 x 10^3 cfu/g	NA	(1-3) x 10^2 -10^3 cfu/g	OK FDA

Preservatives and antioxidants may be put into some food and drug products to make their shelf life longer. Some companies use induction sealing and vacuum pouches to add to the shelf life of their products. Here, in the study the sodium benzoate was added as preservatives to prolong the shelf life of the product, nevertheless the microbiological analysis also play as crucial factors for the food quality and food safety to be consumed by the consumer.

Conclusion

As the conclusion and summary from Table 1, the full analysis shown that the spicy (chilli) soy sauce is safe to consume after two years with regards to the food quality, food safety and also the microbiological analysis not exceeding 10^4 cfu/g.

Acknowledgement

The authors would like to thank Ministry of Education Malaysia (MOE) for sponsoring the Demand-Driven Innovation Project by Public Private Research Network (PPRN) and also to thank Universiti Tenaga Nasional (UNITEN) for giving the permission to publish this paper.

References

[1] Michael, D. P., John, N. S., A.L. Branen, A. L. (2005). Antimicrobials in foods, Third edition, CRC Press, 29-34.
[2] Ogiehor, I. S., Ikenebomeh, M. J. (2005). Extention of shelf life of garri by hygienic handling and sodium benzoate treatment, Volume (4), *African Journal of Biotechnology*, 744-748.
[3] Peter, G. (1995). Traditional soyfoods: processing and products. American Institut of Nutrition, 570.
[4] Pitt, J. I., Hocking, A. D. (2007). Fungi and food spoilage, Third edition, Springer Dordrecht Heidelberg, London, New York, 401-419.
[5[Yadav, P., Garg, N., Kumar, S. (2014). Improved shelf stability of mulberry juice by combination of preservatives, Volume (5), First edition, *Indian Journal of Natural Product and Resources,* 62-66.
[6] http://www.saveur.com/article/Recipes/soy-chili-sauce. Retrieved on 5 October 2016.

Study on Handing Process and Quality Degradation of Oil Palm Fresh Fruit Bunches (FFB)

Zainon Binti Mat Sharif[1,*], Norhasnina Binti Mohd Taib[1], Mohd Sallehuddin Bin Yusof[2], Mohammad Zulafif Bin Rahim[2], Abdul Latif Bin Mohd Tobi[3], Mohd Syafiq Bin Othman[4]

[1]Graduate Business School, College of Graduate Studies, Universiti Tenaga Nasional, Putrajaya Campus, Jalan IKRAM-UNITEN, Kajang, Selangor, Malaysia.

[2]Faculty of Mechanical and Manufacturing Engineering, University Tun Hussein Onn Malaysia (UTHM) P.O. Box 101, Parit Raja, Batu Pahat 86400, Johor, Malaysia.

[3]Faculty of Engineering Technology, University Tun Hussein Onn Malaysia (UTHM) P.O. Box 101, Parit Raja, Batu Pahat 86400, Johor, Malaysia.

[4]Faculty Hotel and Tourism, Universiti Teknologi Mara, Bertam Campus, 13200 Pulau Pinang, Penang, Malaysia.

*zainon@uniten.edu.my

Abstract. The main objective of this study is to determine the relationship between quality of oil palm fresh fruit bunches (FFB) and handling processes. The study employs exploratory and descriptive design, with quantitative approach and purposive sampling using self-administrated questionnaires, were obtained from 30 smallholder respondents from the Southern Region, Peninsular Malaysia. The study reveals that there was a convincing relationship between quality of oil palm fresh fruit bunches (FFB) and handling processes. The main handling process factors influencing quality of oil palm fresh fruit bunches (FFB) were harvesting activity and handling at the plantation area. As a result, it can be deduced that the handling process factors variable explains 82.80% of the variance that reflects the quality of oil palm fresh fruit bunches (FFB). The overall findings reveal that the handling process factors do play a significant role in the quality of oil palm fresh fruit bunches (FFB).

Keywords: Oil palm fresh fruit bunches, handling processes

1. Introduction

The palm oil industry is one of the key economic drivers of the agricultural sector in developing countries such as Malaysia and Indonesia. Both countries have been competing with each other to increase their productivity and improve their yield. Palm oil brings people together, due to its flourishing and profitable business. The smallholders took this opportunity to join this business. According to the Malaysian Palm Oil Board (MPOB), as at December 2015 the number of smallholders have grown almost 47% to 177,046 from 120,437 in 2007; as at January this year, smallholders have exceeded 180,000. Malaysia is a big player in palm oil and it is the fourth largest contributor to the national economy. Malaysia's palm oil industry currently accounts for RM53billion in GNI [13]. Khoo and Chandramohan [7] mentioned palm oil is the second largest oils and fats produced after soybean oil, accounting for 23.6 million tonnes or 20% of the world oils and fats production since 2001. There are two value chains that Malaysia's palm oil spans which is upstream and downstream plantation. Having limited land to expand oil palm plantations, the industry needs to enhance upstream plantation to sustain growth in this sector. The interest in oil palm cultivation in Malaysia remains strong despite issues of land scarcity and limited human capital (Intan Farhana Zainul, [6]). As land is scarce, palm oil and development is working hard to greatly increase the yield without having to use more land. Thus, the introduction of the National Key Economic Area (NKEA) has the potential to contribute a quantifiable amount of economic growth to the Malaysian economy and hence help Malaysia achieve high income by 2020.

2. Background of the Study

2.1 Quality of Oil Palm Fresh Fruit Bunches (FFB)

Quality of oil palm fresh fruit bunches (FFB) is measured by low free fatty acids (FFA) and high oil extraction rate (OER). The smallholders are making efforts to produce high quality oil palm FFB by enhancing the proper way of handling the oil palm FFB so that the oil palm FFB would not be damaged or bruised while the handling process takes place. Moreover, the quality of oil palm FFB also could be differentiated by observing young, ripe, and overripe fruits. The Palm Oil Mill Administrator [18] stated that a ripe fruit bunch is a bunch that has fruit that are reddish orange in colour and has some fruit that have been detached. If the fruits are young, it will contribute none or zero oil component. Bunches with long stalks and very young bunches will undoubtedly reduce the extraction rate (Abdul & Tan [1]). However if the fruits are overripe and ripe these will contribute to high OER where the fruits contain a lot of oil component. The research Abdul and Tan [1] found that in achieving high rates of extraction, bunches have to be in the state of optimal ripeness. Therefore the fruits should be graded correctly by an individual appointed by the company and must have undergone grading training courses. The purpose of grading is to ensure that oil palm FFB is harvested and collected according to the standard of ripeness. Besides that, the aim of this oil palm FFB manual grading is to improve the quality and quantity of crude palm oil and kernel production in Malaysia. The Palm Oil Mill Administrator [18] also stated that the purpose of oil palm FFB grading is to determine the quality of oil palm FFB received and thus helps to determine the purchase price. Whenever the quality of oil palm FFB is high, the smallholders would receive a worthy price according to premium price and premium quality. This will also ensure that the suppliers and millers obtain a fair deal from their transactions.

2.2 Harvesting Activity

The important process in oil palm plantation is to harvest oil palm fresh fruit bunches at the optimum ripeness for maximum oil yields as the content of a fruit bunch is a function of its degree of ripeness. The extraction rate obtained by the factory is greatly affected by oil palm fruit ripeness standards since the oil content of fruit increases rapidly. Bunch colour begins to change from black to reddish-orange until full ripeness is achieved. The most usual method of defining the ripeness standard is by reference to the number of loose or detached fruits, with ten loose fruits on the ground being a commonly-used standard. Oil palm mature has specified a standard that mature bunches must have at least one detached loose fruitlet on the ground at the base of the palm before the bunch is harvested. According to Geoffrey [4], fruits may be damaged in the process of pruning palm fronds to expose the bunch base to facilitate bunch cutting. This shows that if the harvester harvests the fruit roughly it will damage the oil palm FFB. Malaysian Standard 1784 [11] stated that oil palm FFB should be harvested according to acceptable industry ripeness standards. This indicates that grading is important in the harvesting process in order to distinguish which fruits are ripe to be harvested by the harvester. It was also clearly stated that all loose fruits should be collected without contamination by ground debris and stones.

Furthermore, in the harvesting process it is important that the harvester collect all the loose fruit because loose fruits contain more oil component as compared to bunches. Whenever the harvester collects the loose fruits without it being contaminated by soil debris and stones, it will increase the extraction rate. There should also be zero tolerance for unripe bunches as unripe bunches do not have any oil component. Harvesting correctly is crucial and exerts considerable influence on both oil extraction rates and the final crude palm oil (CPO) quality. If smallholders harvest poor quality bunches, it will lead to high FFA and low OER.

2.3 Handling at Plantation Area

Material handling is the general term for moving materials within an organization. Every time those materials are moved around operations it will use materials handling. The aim is for efficient movements with short journeys, using proper equipment with little damage and delay. Material handling is concerned with the movement of goods within that space. One of the definitions adopted by the Materials Handling [12] is that materials handling is the art and science involving the moving, packing, and storing of substances in any form. The quality chain regarding handling starts from the harvesting process. After the harvester have harvested the oil palm FFB, they should handle the oil palm FFB with care because it could easily become bruised and damaged if it is not handled properly. Minimal post-harvest handling and damage to oil palm FFB is important to maintain an acceptable FFA content in the crude palm oil. During the handling process at the plantation area, it is vital for the harvester to collect all the loose fruits that fall from the oil palm. The oil palm mature specified that the complete collection of loose fruits is important because they contain up to 48% oil while bunches contain approximately 22% oil. In addition, the equipment used during the handling process in the plantation area is also important. As we know, handling consists of moving the material from one place to another, thus, equipment such as chisel, sickle, small axe, and wheelbarrow are needed while handling the oil palm FFB at the plantation area. Harvester efficiency will reduce when poor quality wheelbarrows are used for harvesting and infield collection. Other than that, all ripe bunches and all loose fruits must be harvested, stacked neatly at the ramp or roadside to be collected and delivered to the intermediate collection center which is the fruit dealer. All oil palm FFB and loose fruits collected from the palm base should be delivered to the collection points with minimal damage, delay and contamination, as per the Malaysian Standard 1784 [11].

2.4 Handling at Intermediate Collection Center

At the intermediate collection center, it is important to minimize fruit damage because it will then be transported to the mill where the oil palm FFB will be sterilized and processed to produce a good quality of crude palm oil. For example, at the ramp during loading and unloading of the oil palm FFB where it will be graded to distinguish between the good and damaged fruits so as to maintain the quality of the crude palm oil. Malaysian Standard 1784 [11] stated that oil palm FFB should not be shoveled into vehicles from the intermediate ramp floor to minimize bruising. This indicates that the bruises on the fruit will contribute to fruit damage and hence will increase the FFA. When the FFA is increased, the extraction rate will consequently decrease. Geoffrey [4] mentioned that research has shown that if the fruit is bruised, the FFA in the damaged part of the fruit increases rapidly to 60 percent in an hour. In other words, the oil palm FFB should be transported to the mill within 24 hours after being harvested. If transport of the oil palm FFB to the mill was delayed, it will affect the oil palm FFB quality which will cause infections. As per Abdul and Tan [1], the delay in transferring harvested bunches especially the overripe ones, allows these bunches to be infected by microorganisms thereby resulting in rotting fruits. A study made by Abdul and Tan [1] identified that the state of deterioration has been attributed mainly to handling and transportation of oil palm FFB to the mill and also the condition of the oil palm FFB at the collecting platform and ramp. This shows that the oil palm FFB is prone to damages and bruises which then will affect the quality of the oil palm FFB during the handling process and transportation of oil palm FFB whether at the collecting center or ramp.

2.5 Oil Palm FFB Protection during Transportation

The vehicles must be in clean condition in order to preserve the quality of the oil palm FFB so that the oil palm FFB will not be contaminated by the residues left on the vehicles. Based on MPOB [10], the compartments should be dedicated to oil palm FFB, free from previous load residues, and free from odors of previous loads. The driver of each transport should check the vehicle before each consignment is loaded into the compartment. Furthermore, the vehicles have to keep the oil palm FFB covered at all times while the transportation of oil palm FFB is in progress. This is to ensure that the oil palm FFB is prevented from being exposed to rainwater or bird droppings. If the oil palm FFB gets wet, the water will penetrate into the fruits and thus will affect the quality whereas if the fruits were exposed to bird droppings, it will lead to contamination. MPOB [11] also states that the penetration of rainwater and splash water should be prevented during transport. Besides that, the vehicles should be free from carrying any hazardous cargo to avoid any damages during the transportation. Hazardous cargo such as materials that are radioactive, flammable, explosive, corrosive, toxic and many more is not allowed on the vehicles because it could affect the quality of the oil palm FFB. According to Malaysian Standard 1784 [11], the vehicles transporting oil palm FFB shall be registered and licensed and secured and should not carry other hazardous cargo e.g. chemicals. When the vehicles arrive at the intermediate collection center, it must be weighted at the ramp and then recorded on the proper documentation so that they could keep track of the oil palm FFB to prevent loss and pilferage during the conveyance as this is to retain the quality of the oil palm FFB. As per MPOB (2008), in the case of delivery to third parties, all consignment of oil palm FFB for transport to the purchaser must be accompanied by proper documents which must meet the minimum regulatory requirements set by the MPOB.

The palm oil industry is the fourth largest contributor to the Malaysian Gross National Income (GNI) (NKEA, [9]). Thus, referring to the Prime Minister's program which is NKEA, the EPP2 aims to increase the national oil palm FFB yield from 21 tonnes per hectare to 26.2 tonnes per hectare by 2020 (MPOB, [13]). In order to increase the yield the oil palm FFB producers need to supply good quality oil palm FFB to the market and to do this, it involves the handling process of the oil palm FFB. As quality of goods is very important to meet consumer demand, it is crucial to take into account how the raw materials are handled. Material handling is important to ensure movement of goods is in a right quantity and in good quality. Tuan Hj. Shamsudin Selamat the Chairman of the Intermediary Collection Center did mention that "Generally, Malaysia has a similar type and condition of land and weather. Unfortunately, the oil palm FFB quality is not similar. This reflects the situation of something wrong somewhere". From that statement there may be several problems concerning quality of oil palm FFB.

The problems that might be faced by the company which affect oil palm FFB quality regarding their handling process are:

i) Harvesting activity
   In order to harvest the oil palm FFB, the harvesters have to prune the fronds. The oil palm FFB will be harvested by the harvester and they need good skills to have a better understanding of how to harvest the oil palm FFB, identify the oil palm FFB correctly whether overripe, ripe, or unripe so that they harvest the oil palm FFB that is in the state of optimal ripeness.

ii) Handling at plantation area
   This refers to the handling process of raw materials from the plantation area to ramp where the oil palm FFB will be placed and assembled before being carried to the collection premise.

iii) Handling at intermediate collection center
When the raw materials reach the collection premise, there will be another handling process that took place. The weight of the oil palm FFB will be measured and recorded and then the oil palm FFB will be placed and assembled before being transported out to the mill.

iv) Oil palm FFB protection during transportation
This refers to steps taken for protection during the transportation of oil palm FFB from the plantation area to the mill. How the drivers preserve the freshness of oil palm FFB when transporting it to other locations. Example of vehicles used during the transportation is a small lorry that has capacity of 1 tonnes and a big lorry that has capacity of 30 tonnes.

The main objective of this study is to determine the relationship between quality of oil palm fresh fruit bunches (FFB) and handling processes.

3. Methodology

Figure 1 shows the framework of the research.

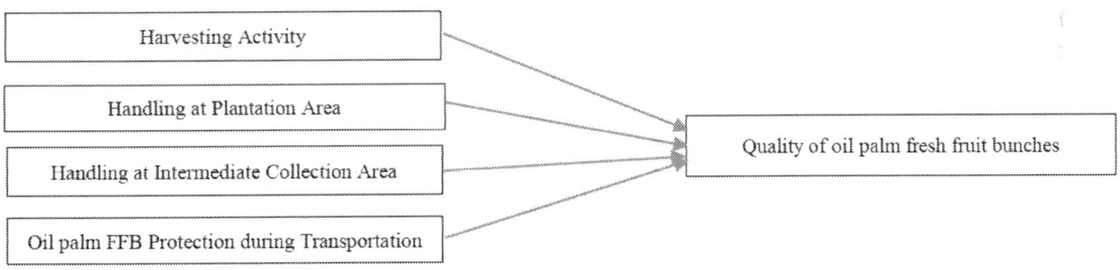

Figure 1: Research Framework on the relationship between Quality of oil palm FFB and Handling Process

The study employs exploratory and descriptive design, with quantitative approach and purposive sampling using self-administrated questionnaires. The sample size was chosen among smallholders in the Southern Region, Peninsular Malaysia. Lansford B. R & Lansford T. R [8] suggested that the minimum number of sample size could be 28. The instrument used in the questionnaires was adopted from Zainon et. al., [19, 20, 21, 22, and 23].

4. Results and Discussion

Table 1: Respondents' Demographic Profile

Item		Frequency	Percentage (%)
Gender	Male	25	83.8
	Female	5	16.7
Area of Plantation	< 1 acre	0	0
	1-2 acre	22	73.7
	> 2 acre	8	26.3
Age of Oil Palm	< 3 years old	0	0
	3-10 years old	21	73.3
	> 10 years	9	26.7

	old		
Monthly Harvesting Frequency	Once	0	-
	Twice	30	100
	More	0	-

Table 2: Reliability Analysis

Variables	No of item	*p or* (α)	M	SD
Quality of oil palm fresh fruit bunches (DV)	3	0.884	3.778	0.813
Harvesting activity (IV_1)	5	0.844	3.213	0.934
Handling at plantation area (IV_2)	5	0.848	3.527	0.744
Handling at intermediary collection centre (IV_3)	5	0.873	3.720	0.798
Oil palm fresh fruit bunches protection during transportation (IV_4)	5	0.853	3.727	0.723

Note: M – mean, SD –Standard Deviation

Table 3: Pearson Correlation Matrix

Variable	DV	IV_1	IV_2	IV_3	IV_4
DV	1.000				
IV_1	0.827^{**}	1.000			
IV_2	0.808^{**}	0.709^{**}	1.000		
IV_3	0.521^{**}	0.574^{**}	0.604^{**}	1.000	
IV_4	0.434^{**}	0.304	0.672^{**}	0.690^{**}	1.000

Note ** Correlation is significant at the 0.01 level (two-tailed)
* Correlation is significant at the 0.05 level (two-tailed)

Table 4: Model Summary

Model	R	R Square	Adjusted R Square	Std. Error of the Estimate
1	0.910^a	0.828	0.800	0.36368

Table 5: ANOVA

Model	Sum Square	Df	Mean Square	F-test	Sig.
Regression	15.879	4	3.970	30.013	0.000^b
Residual	3.307	25	0.132		
Total	19.185	29			

Note: ** Correlation is significant at the 0.01 level (two-tailed)
a. Dependent Variable: Quality of Oil Palm FFB (DV)
b. Predictors (constant): IV_1, IV_2, IV_3, IV_4

This study was conducted in the southern region of Peninsular Malaysia. Based on the respondent profile, 83.8% of the respondents are male and the remaining 16.7% are female. They are considered as small farmers since 73.7% or 22 of the respondents have 1 to 2 acres of palm oil plantation and only 8 respondents (26.3%) have more than 2 acres of palm oil plantation. All of them harvest their palm oil fruits twice a month, as shown in table 1. All variables are having a good α value for Cronbach alpha (α) test. According to Tabachick [17], an alpha (α) value above 0.8 and less than 0.9 are considered good. The quality of oil palm FFB having the highest α value at p = .884 followed by handling at intermediary collection center with $p = 0.873$ and harvesting activity having the lowest α at $p = 0.844$, as shown in table 2. Pearson correlation analysis has found that all independent variables correlate significantly with the dependent variable as shown in table 3. The results show that, all independent variables in this study correlates significantly with the dependent variable which is the quality of the fresh fruit bunch. Nevertheless, according to Guildford Rule of Thumb, among the independent variables the harvesting activity has the highest correlation with $p = 0.827$. This result confirmed that the harvesting activity is the most important process in ensuring the quality of the oil palm fruits. The harvester has to have enough knowledge in determining the level of optimum ripeness to gain maximum oil extraction and final crude palm oil quality. Harvesting premature fruits will lead to low quality of output and wasting of resources in the production. Handling at the plantation area is found to be the second important factor in determining its quality. Results show in table 3, that handling at the plantation area correlate significantly with $p = 0.808$. Where handling at intermediary collection center and protection during transportation are also significant at $p = 0.521$ and $p = 0.434$ respectively. Having the optimum oil palm FFB alone does not ensure the quality if it is not properly handled before the oil palm FFB reaches the extraction process. The time taken to the processing center, the equipment used in handling the oil palm FFB and probably the infrastructure (i.e. transportation and road condition) may reduce the fruits damages and thus contribute to the quality of the oil palm FFB. As a result, harvesting activity, handling at plantation area, handling at intermediate collection center, and oil palm FFB protection during transportation as independent variables with significance can explains 82.80% (R square) of the variance that reflects the quality of oil palm fresh fruit bunches (FFB) as dependent variable as shown in table 4. The pilot test study model was a fit model according to ANOVA table 5, with the significant value $p < 0.05$.

5. Conclusion

In order to maintain the oil palm FFB quality, harvesting activity should be given priority especially in selecting which fruit bunch is ready for harvest. Other than that, handling process and fruit protection after harvesting also need to be considered to ensure the quality of oil palm FFB. By having good quality oil palm FFB, it will help increases the production of the crude oil palm per ton as land scarcity is our main issue in expanding the industry. The limitation of the study employs exploratory and descriptive design, with quantitative approach and purposive sampling using self-administrated questionnaires, were obtained from 30 smallholder respondents from the Southern Region, Peninsular Malaysia. The recommendation of the study for the future researcher is to conduct an actual sampling based on the actual population of smallholders in Malaysia and to employ this pilot study test model.

Acknowledgement

The authors would like to express greatest appreciation to the Dean of College of Graduate Studies (COGS), and Universiti Tenaga Nasional (UNITEN) for giving the permission to publish this paper.

References

[1] Abdul, A. A., & Tan, Y. A. (1989).*The effects of handling of OIL PALM FFB on the formation of FFA and the subsequent quality of crude palm oil*. In J. Sukaimi (Ed.) Proceedings Palm oil Development Conference, pp. 482-485. Kuala Lumpur: PORIM

[2] Agriculture and Consumer Protection. (n.d). *Small-scale palm oil processing in Africa*, Retrieved on 2nd October 2013 at 11.00 am from http://www.fao.org/docrep/005/y4355e/y4355e04.htm#TopOfPage

[3] Chong Chiew Let. (1995). *Handling Storage and Transportation of Fresh Fruit Bunches and Palm Oil Products.* Palm Oil Technical Bulletin. Kuala Lumpur: MPOB

[4] Geoffrey (2002).*Small-scale palm oil processing in Africa Issue 148*.KwasiPokin. Food & Agriculture Organization of the United Nation

[5] George, D & Mallery, P. (2001).*SPSS for windows steps by steps: A simple guide and reference, 10.0 update*. Third edition, USA, Allyn& Bacon.

[6] Intan Farhana Zainul (2013) *Analysts see huge potential in the palm oil industry for small-time players*. Retrieved on 26th September 2013 from http://www.thestar.com.my/

[7] Khoo, K. M., & Chandramohan, M. (2002).Malaysian Palm Oil Industry at Crossroads and its Future Direction. *Oil Palm Industry Economic Journal (Vol.2 (2)/2002) p.10-15*

[8] Lunsford, B. R., & Lunsford, T. R. (1995). Research Forum - the research sample part II: Sample size. *JPO, 7(4), 137-141.*

[9] NKEA (2011).*National Key Economic Area*, Ministry of Plantation Industries & Commodities, Malaysia 2011

[10] Malaysian Palm Oil Board. (2008). *Code of Good Practice for the Handling, Transport and Storage of Products from the Oil Palm*. Kuala Lumpur: MPOB

[11] Malaysian Standard 1784. (2006). *Good Agricultural Practice Part 2: Oil Palm*. Kuala Lumpur: Department of Standards Malaysia.

[12] Materials Handling, MHI (n.d) retrieved on 2nd October 2014 at 11:40 am from http://discovery.bits-pilani.ac.in/dlpd/courses/coursecontent/courseMaterial/mmzg511/MOML12.pdf

[13] MPOB (2015) *Overview of the Malaysian Palm Oil Industry*. Retrieved on 2 March 2016 from www.mpob.gov.my

[14] MPOC (2012) *Malaysian palm oil industry*. Retrieved on 24th September 2013 from http://www.mpoc.org.my/Malaysian_Palm_Oil_Industry.aspx 2012

[15] Oil Palm Mature (n.d) Retrieved on 3rd October 2015 at 2:17 pm from http://www.ipni.net/ppiweb/filelib.nsf/0/9BF9AC573FE5E3DB48256B49002F64E7/$file/OP%20HB%20Mat%20p1-8.pdf

[16] Sekaran, U., & Bougie, R. (2012).*Research methods for business*. (5th ed.). United Kingdom: John Wiley & Sons, Ltd.

[17] Tabachnick, B. G. (2015). *Using Multivariate Statistics 6th Edition*. California State University.

[18] The Palm Oil Mill Administrator (2015) *Fresh Fruit Reception*, Retrieved on 10thOctober 2015 at 9.00 am from http://palmoilmill.co.nz/index.php?option=com_content&view=article&id=47:fresh-fruit-reception&catid=47:oil palm FFB-reception&Itemid=71

[19] Zainon M. S. (2009a). *Learner Motivation, Training-related Factors, Work Environment and Learning as Determinants of Transfer of Learning to the Workplace.* Unpublished Ph.D. Thesis, Universiti Putra Malaysia.

[20] Zainon M. S., Aminah A., Rahim M. S., and Jamilah O. (2009b).*Individual, Situational and Training Design Factors as Determinants of Transfer of Learning in the Workplace.* Proceedings International Conference on Agriculture Extension 2008 AGREX08 16-19 June 2008 at Equatorial Hotel, Bangi-Putrajaya by Universiti Putra Malaysia, Serdang. 2009. Pg 164-172.

[21] Zainon, M. S., S.Nurulakla, S. A., and M.Yusof.,M (2011). The Effects of Learners Motivation, and Learning on Transfer of Learning to the Workplace in Malaysia. *Journal of Management and Science, Volume 9, Number 2, December 2011, pg 83 – 92*, Management and Science University (MSU), Shah Alam, Selangor, Malaysia.

[22] Zainon, M. S., S.Nurulakla, S. A., and M.Yusof.,M. (2012a). Empirical Evidence on Learning and Transfer of Learning. *Journal of Management and Science, Volume 10, Number 1, June 2012, pg 33 – 45, Management and Science University (MSU), Shah Alam, Selangor, Malaysia.*

[23] Zainon, M. S., S.Nurulakla, S. A., and M.Yusof.,M. (2012b). Training-related Factors Play an Important Role on Transfer of Learning. *Journal Human Resource Development, Burapha University, Thailand, Vol.3, No.1, June 2012, pg. 35 – 49, ISSN: 1906-9308*

Institute of Physics
Dirac House, Temple Back
Bristol BS1 6BE UK

ISSN: 1757-8981
ISBN 978-1-5108-4181-9

CURRAN ASSOCIATES INC.
proceedings.com